Tourism and Recreation

Stephen Williams

An imprint of Pearson Education

Harlow, England • London • New York • Boston • San Francisco • Toronto • Sydney • Singapore • Hong Kong
Tokyo • Seoul • Taipei • New Delhi • Cape Town • Madrid • Mexico City • Amsterdam • Munich • Paris • Milan

Pearson Education Limited
Edinburgh Gate
Harlow
Essex CM20 2JE
England

and Associated Companies throughout the world

Visit us on the World Wide Web at:
www.pearsoneduc.com

First published 2003

© Pearson Education Limited 2003

The right of Stephen Williams to be identified as author of this work has been asserted by the author in accordance with the Copyright, Designs and Patents Act 1988.

All rights reserved. No part of this publication may be reproduced, stored in a retrieval system, or transmitted in any form or by any means, electronic, mechanical, photocopying, recording or otherwise, without either the prior written permission of the publishers or a licence permitting restricted copying in the United Kingdom issued by the Copyright Licensing Agency Ltd, 90 Tottenham Court Road, London W1T 4LP.

ISBN 0 582 32026 7

British Library Cataloguing-in-Publication Data
A catalogue record for this book is available from the British Library.

10 9 8 7 6 5 4 3 2 1
08 07 06 05 04 03

Typeset in 10/12pt Sabon by 35
Printed in Malaysia, PA

The publisher's policy is to use paper manufactured from sustainable forests.

Tourism and Recreation

Themes in Tourism

Series Editor: Professor Stephen J. Page, *Scottish Enterprise Forth Valley Chair in Tourism, Department of Marketing, University of Stirling, Stirling, Scotland FK9 4LA.*

The Themes in Tourism Series is an upper level series of texts written by established academics in the field of tourism studies and provides a comprehensive introduction to each area of study. The series develops both theoretical and conceptual issues and a range of case studies to illustrate key principles and arguments related to the development, organisation and management of tourism in different contexts. All the authors introduce a range of global examples to develop an international dimension to their book and an extensive bibliography is complemented by further reading and questions for discussion at the end of each chapter.

Books published in the Themes in Tourism Series

S.J. Page *Transport and Tourism*
C.M. Hall *Tourism Planning*
S.J. Page and R.K. Dowling *Ecotourism*
R. Scheyvens *Tourism for Development*
S.J. Page and C.M. Hall *Managing Urban Tourism*
D.J. Timothy and S.W. Boyd *Heritage Tourism*

Contents

List of figures		*vii*
List of tables		*ix*
Preface		*xi*
Acknowledgements		*xiii*
Publisher's acknowledgements		*xiv*

1 Tourism, recreation and leisure — 1

    Introduction — 1
    Tourism, recreation and leisure: the problems of definition — 2
    Tourism and recreation: finding the common ground — 9
    Recreation and tourism: convergence and de-differentiation — 20
    The structure of the book — 23
    Questions — 24
    Further reading — 24

2 Seaside resorts as centres of tourism and recreation — 25

    Introduction — 25
    Spas, sea bathing and the growth of coastal resorts — 26
    Resorts as tourist places — 36
    Tourism, recreation and the seaside – post-1945 developments — 41
    Conclusion — 56
    Questions — 56
    Further reading — 56

3 Tourism, recreation and international travel — 58

    Introduction — 58
    The growth of international tourism since 1945 — 59

	Factors promoting global tourism	71
	Convergence between recreation and foreign travel	86
	Questions	88
	Further reading	88
4	**Tourism and recreation in urban places**	**89**
	Introduction	89
	Urban recreation and tourism: place, space and activity	91
	Changing patterns in urban recreation	106
	New urban places of recreation and tourism	112
	Conclusion	122
	Questions	123
	Further reading	123
5	**Tourism and recreation in the countryside**	**124**
	Introduction	124
	Evolution of recreation and tourism in the countryside to 1945	125
	Patterns of rural recreation and tourism since 1945	131
	Conclusion	152
	Questions	153
	Further reading	153
6	**Tourism and recreation: issues and policy approaches**	**155**
	Introduction	155
	Recreation and tourism: key policy issues	156
	Recreation and tourism: responses to policy issues	166
	Recreation and tourism: policy in practice	179
	Conclusion	189
	Questions	190
	Further reading	190

Bibliography   *191*
Index   *207*

List of figures

1.1	Tourism or recreation? Holidaymakers and day visitors pack the beaches at Brighton	5
1.2	The structure of demand in travel and tourism	6
1.3	The relationship between leisure, recreation and tourism	7
1.4	The nature of the tourist experience	15
1.5	The nature of the recreational experience	16
1.6	Converted dockland in Gloucester provides a new set of leisure sites based upon heritage museums, restaurants, bars and shopping arcades	19
2.1	The Brighton Pavilion – built for the Prince Regent (later King George IV)	29
2.2	Major seaside resorts in Germany, circa 1900	30
2.3	The spread of seaside resorts in England and Wales, 1750–1911	35
2.4	The beach, the hotel and entertainment – three key elements in the Victorian–Edwardian middle-class seaside holiday (Llandudno, North Wales)	40
2.5	Changing levels in the main mode of holiday transport in the UK, 1960–98	46
2.6	Day excursion destinations served by coach operators in Torbay, south-west England	47
2.7	The historic naval town of Dartmouth (south Devon), a popular attraction for visitors from neighbouring resort areas in Torbay	48
2.8	Seasonal pattern of holiday-taking in the UK, 1999	50
2.9	Distribution of holiday camps in England and Wales, 1939	52
2.10	Part of the redevelopment of St Katherine's Dock, London	54
2.11	Recreational and tourist development at South Dock, Swansea	54
2.12	Port Solent Marina development, Portsmouth	55
3.1	Growth of international tourist arrivals, 1950–99	59
3.2	Actual and projected growth in cross-Channel traffic, 1994–2003	64
3.3	Growth in international tourism in major market areas, 1986–96	68

3.4	Pattern of outbound international tourism from Europe to WTO regions, 1996	69
3.5	Expansion of the British air charter market, 1963–72	77
3.6	Patterns of multiple holiday-taking by UK residents, 1972–96	81
3.7	The European ski holiday market, 1998	84
4.1	Tourist and recreational developments around Centenary Square, Birmingham	94
4.2	Part of Birmingham's Centenary Square–Brindley Plaza redevelopment	95
4.3	The structure of the urban tourist product	96
4.4	Visitor levels at major tourist attractions in central Paris	97
4.5	Tourist attractions and facilities on the Île de la Cité, Paris	99
4.6	Functional areas and theoretical linkages in the tourist city	100
4.7	The traditional image of the urban park – ornamental horticulture, seating and parkland sports (Stafford)	110
4.8	Part of Telford town park (Shropshire)	112
4.9	Theoretical relationships between shopping, leisure and place	113
4.10	The modern shopping mall as a leisure environment	116
4.11	Primary land uses in the Mall of America, Minneapolis (ground floor)	118
5.1	Picturesque landscapes (Loch Creran, Scotland)	128
5.2	Origins of visitors to Dartmoor National Park	133
5.3	Profile of day visitors to urban, rural and coastal destinations in the UK	135
5.4	Monthly pattern of visits to the countryside in the UK	136
5.5	Contrasting temporal patterns of countryside visiting in the UK	137
5.6	Distances travelled on countryside trips in the UK	137
5.7	Locations and levels of visiting to major theme parks in England and Wales, 1999	140
5.8	The proposed EuroVelo cycle network in Europe	142
5.9	Visitor attractions at Amerton Farm, Staffordshire	146
5.10	Integration of recreation and tourism at Bewl Bridge reservoir, Sussex	147
6.1	Positive impacts associated with heritage developments in Liverpool	165
6.2	Designated land in England and Wales	171
6.3	Major areas of inland water-based recreation in Wales	174
6.4	Compatibility matrix for water-based recreation	175
6.5	Zoning strategy in the Peak District National Park	176
6.6	Zoning pattern at Kingsbury Water Park, Warwickshire	177
6.7	Part of the interpretive displays at the High Moorland Visitor Centre, Dartmoor National Park	179
6.8	The redevelopment of Birmingham's Gas Street Basin	181
6.9	Visitor attractions and leisure areas in Manchester city centre	183

List of tables

1.1	Comparison of motives for participation in leisure/recreation and tourism	13
2.1	The concentration of the elderly in coastal districts containing resorts – southern England, 1991	44
2.2	Level and frequency of holidaymaking by UK residents (4 or more nights away), 1971–98	50
2.3	Accommodation used on main holidays in Great Britain, 1955–98	53
3.1	International tourist arrivals by WTO region, 1986–96	60
3.2	Changing levels of international arrivals in selected eastern European states, 1992–6	62
3.3	Foreign hotel nights in European cities, 1992	66
3.4	Attendance at selected attractions in Europe, 1995	67
3.5	Market shares of long-haul and intra-regional tourism in WTO regions, 1996	70
3.6	Changing structure of the East Asia and Pacific region market, 1985–96	70
3.7	Growth of inclusive tours markets in Britain, Germany and Sweden, 1988–96	73
3.8	Variations in mode of transport used by European holidaymakers	75
3.9	Changes in the socio-economic structure of Great Britain, 1980–2000	80
3.10	Recent changes in location of holidays taken by British tourists	81
3.11	Dimensions of 'aesthetic cosmopolitanism'	82
4.1	European city holiday trips – major destinations, 1995	92
4.2	Distribution of recreational facilities and spaces within intra-urban zones – a model pattern	103
4.3	Principal recreational interests and activities in the UK, 1997–9	105
4.4	Typical attributes of visitors to 10 urban parks in the UK	110

5.1	Frequency of countryside visiting and the share of visits in the UK, 1990	138
5.2	The continuing appeal of traditional countryside activities amongst visitors to national parks in the UK	139
5.3	Overnight stops by cycle tourists on part of the Danube Cycle Route (*Donauradweg*), Austria, 1994	143
5.4	A typological framework for farm-based recreation	151
6.1	Patterns of landownership in national parks in England and Wales, 1990	159
6.2	Principles for tourism development in the British countryside	163

PREFACE

At the start of the twenty-first century, tourism has become a phenomenon of global significance. Every year millions of people travel as international tourists and billions make domestic tourist trips. As an industry (or collection of industries) it creates employment and sustains economies at local, regional and, occasionally national levels. It may be a force for social and cultural change and in many situations tourism creates significant impacts upon the environment too. Although there are important sectors of tourism that relate to the activities of business or purposeful travel for reasons of health, religion, or education, the largest element in most tourism markets is travel for pleasure. In this way, tourism becomes a part of the much larger realms of leisure and recreation, yet as this book will try to argue, the nature and character of the relationship between tourism and these wider areas has been subject to a surprising degree of neglect.

The seeds of neglect are rooted in the early origins of the study of tourism as an academic subject. I first became interested in tourism and recreation as a graduate student working in the University of Wales. At that stage (the early 1970s) the academic study of the field was very much in its infancy. In north America, important pioneering work had been done by the Outdoor Recreation Resources Review Commission (ORRRC) and writers such as Marion Clawson and Thomas Burton had already published some influential work. In the UK, Allan Patmore's ground-breaking study of *Land and Leisure* was published in 1970 and quickly became instrumental in stimulating both teaching and research in the field. Yet even at this formative stage, the traditions of separate study of tourism and recreation were discernable and much of the subsequent development (within what has become a burgeoning field of study) has often tended to perpetuate the notion of parallel, but rarely intersecting, investigation into the nature of tourism and recreation. This tradition has been sustained – in many areas of teaching, in research, in the organisation of conferences and in the workplace – even though tourism and recreation are widely understood to be related phenomena. Consequently, the nature of that relationship has been seldom examined.

Textbooks normally seek to work in several related ways, so that whilst the following chapters are intended to have a value as an up-to-date synthesis and exploration of aspects of the current understanding of recreation and tourism, the book is also an attempt to use that discussion to focus a part of our thinking onto the relationship between tourism and recreation. Given the comparatively small canvas on which I was invited to work, the text – of necessity – represents no more than a series of outline ideas on how tourism and recreation relate to each other, and makes no claims to provide a definitive perspective – if such a thing is attainable in these postmodern times! It is hoped, however, that the discussions that are presented will stimulate the reader to think more deeply and more widely about the notion of relationships in recreation and tourism. By tradition, tourism has been represented in terms of escape from routine, of contrast, and of difference from the kind of motives and behaviours that often typify day-to-day recreations. Yet the more I read and think about the subject, the more I am beginning to question ideas of difference and to recognise that much of the nature of tourism and recreation is strongly reflective of a mutual inter-dependence that has often gone unnoticed in much of the previous literature. Moreover, not only is this evident in the processes of social and economic change that are associated with the emergence of post-industrial patterns and lifestyles – especially in the developed nations of the world – but it is also deeply embedded in the historic processes of tourism development.

Although the book is written by a geographer, I would not wish to present the work as a geography of recreation and tourism (for there is much more to the geography of recreation and tourism than is presented here). However, since it is a work that is written essentially from a geographical perspective, the articulation of ideas and arguments tends to reflect the geographer's traditional concerns for people and place, for spaces and patterns, and for change through time. Hence the text revisits some familiar territory: the development of seaside resorts; the growth of international travel; urban and rural patterns of recreation and tourism; and policy issues, as settings in which relationships might be explored. But in so doing, it is hoped that it will reveal new perspectives upon important elements in the environments of recreation and tourism.

Stephen Williams
Staffordshire University, UK

Acknowledgements

The successful completion of a text seldom depends upon the efforts of the author alone and I am happy to acknowledge the help and support of a number of individuals and organisations in the production of this book.

First, I am grateful to the Series Editor, Professor Stephen Page, for the initial invitation to contribute to the Themes in Tourism series and for the subsequent support that he has provided. I would also like to acknowledge the help and support provided by the staff at Pearson Education, in particular Matthew Smith (who was the original commissioning editor), and Morten Fuglevand and Nicola Chilvers who have overseen the final production of the book.

The task of writing was also helped greatly by the active support and encouragement of colleagues in the Department of Geography at Staffordshire University. The latter stages of writing were assisted by the award of a period of sabbatical leave by the School of Sciences and I am particularly grateful to Dr John Ambrose and Dr Louise Bonner who cheerfully took up some of my teaching during my absence. Although he is now retired from the Department, I must also mention the considerable contribution of Professor George Kay. I have enjoyed George's company – as both a colleague and friend – for nearly twenty-five years and the many lengthy discussions that we have had on the nature of tourism and recreation have, I am sure, been highly influential in shaping my understanding of the field of study.

Lastly, I must thank my wife, Jane, who brought her considerable skills as a cartographer to the design of all of the maps and diagrams within the book.

Publisher's acknowledgements

We are grateful to the following for permission to reproduce copyright material:

Figure 1.3 from *The Geography of Tourism and Recreation*, Routledge (Hall, C.M. and Page, S.J., 1999); Figure 1.5 from 'Paths for whom? Countryside access for recreational walking,' *Leisure Studies*, Vol. 15 No. 3, Taylor & Francis Ltd (Kay, G. and Moxham, N., 1996), http://www.tandf.co.uk/journals; Figure 2.3 from *An Historical Geography of Recreation and Tourism in the Western World, 1540–1940*, John Wiley (Towner, J., 1996) © John Wiley & Sons Limited. Reproduced with permission. Table 3.3 from *Tourism in Major Cities*, International Thomson Business (Law, C., 1996); Table 3.4 from 'Attendance trends at Europe's leisure attractions,' *Travel and Tourism Analyst*, 4, EIU (Jenner, P. and Smith, C., 1996); Table 3.7 from 'The package holiday market in Europe,' *Travel and Tourism Analyst*, 4, EIU (Bray, R., 1996); Table 3.8 from *Transport and Tourism*, Addison-Wesley Longman, (Page, S.J., 1999); Table 3.9 from 'Self-catering holidays abroad,' *Leisure Intelligence*, November (Mintel, 1997) © TGI-BMRB International; Table 3.11 adapted from *Consuming Places*, Routledge (Urry, J., 1995); Table 4.1 from 'Urban tourism in Europe,' *Travel and Tourism Analyst*, 6, EIU (Cockerell, N., 1997); Figures 4.4 and 4.5 reprinted from *Tourism Management*, Vol. 19 No. 1, Pearce, D.G., 'Tourist districts in Paris: structure and functions,' pp. 49–65, copyright 1998, with permission of Elsevier Science; Figure 4.9 from 'Shopping and leisure: implications of West Edmonton Mall for leisure and for leisure research,' *The Canadian Geographer*, Vol. 35 No. 3, The Canadian Association of Geographers (Jackson, E.L., 1991); Table 5.1 from 'Detecting patterns of countryside recreation,' *Staffordshire University Department of Geography Occasional Papers*, New Series No. 8 (Kay, G., 1996); Table 5.2 based on information from *Visitors to National Parks: Summary of the 1994 Survey Findings*, Countryside Agency (Countryside Commission, 1996); Table 5.3 and Figure 5.8 from 'Cycle tourism in Europe: EuroVelo,' *Insights*, Vol. 11 Issue A, pp. 143–56, English Tourism Council (Lumsdon, L., 2000); Figure 6.1 from 'Museums, galleries, tourism and regeneration: some experiences from Liverpool,' *Built Environment*, Vol. 26 No. 2, pp. 152–63, Alexandrine Press (Couch, C. and Farr, S., 2000); Table 6.1 from *The National Park Authority: Purposes, powers and administration*, Countryside Agency (Countryside Commission, 1993); Table 6.2 from *Principles for Tourism in the Countryside*, Countryside Agency (Countryside Commission, 1993); Figures 6.3 and 6.6 from *Coastal Recreation Management*, E & FN Spon (Goodhead, T. and Johnson, D. (eds), 1996); Figure 6.4 from *Recreation and Resources: Leisure patterns and leisure places*, Blackwell (Patmore, J.A., 1983); Figure 6.9 from 'Regenerating the city centre through leisure and tourism,' *Built Environment*, Vol. 26 No. 2, pp. 117–29, Alexandrine Press (Law, C.M., 2000). All photographs were taken by the author.

Elsevier Science Limited for an extract from 'Tourist districts in Paris: structure and functions' by D.G. Pearce published in *Tourism Management*, Vol. 19 No. 1 1998; English Tourism Council for an extract from 'Cycle tourism in Europe: Euro velo' by L. Lumsdon published in *Insights*, Vol. 11 March 2000 A-143; John Wiley & Sons Limited for an extract from 'Farm tourism in New Zealand' by M. Oppermann published in *Tourism and Recreation in Rural Areas* ed. R. Butler *et al.*; Kogan Page Limited for an extract from 'Regenerating the city centre through leisure and tourism' by C. Law published in *Built Environment* 26(2) 2000.

In some instances we have been unable to trace the owners of copyright material, and we would appreciate any information that would enable us to do so.

CHAPTER 1

Tourism, recreation and leisure

Introduction

This book is about relationships – in particular, it is about relationships that exist between the leisure areas of recreation and tourism. Outwardly this might appear to be a rather unproductive, even an unoriginal, point of departure – one that has surely been aired repeatedly within the millions of words that have been expended in academic discussion of these two areas over the last 30 years or so. Yet, whilst our understanding of recreation and tourism is now sustained by a rich seam of knowledge, research and scholarship, it is an understanding that has been shaped largely by academic approaches that have maintained a surprisingly steadfast tradition of separate lines of enquiry into these two fields. For the student engaged in the study of recreation and tourism, therefore, the received wisdom is likely to be one that although acknowledging that these two phenomena are related, will be likely to promote the impression that they are fundamentally different.

This book sets out to question this position. It aims to examine similarity rather than difference; to explore those areas where tourism and recreation are actually related rather than free-standing; and to argue that the relationship is much closer than many would imagine. Implicit in this approach is the assumption that by exposing the extent and character of the relationship, we may better understand the true nature of the subject.

The book is written primarily as a geographically informed synthesis, but given the nature of human geography and its primary concerns for associations between people, place and time, other perspectives are drawn quite naturally into the discussion. Hence, part of the argument is that the proximity of recreation and tourism is not merely a contemporary pattern, but is deeply rooted in the history of leisure. Equally, comprehension of the shared meanings and values that people ascribe to recreation and tourism necessitates some engagement with the literature of sociology and cultural studies, whilst evaluation of the ways recreational and tourism impacts are managed, or in which policy is drawing recreation and tourism together, is informed by an understanding of contemporary economics, planning and politics. So whilst in one sense the book is a 'geography' of recreation and tourism, it aims to draw upon a broader base of disciplinary perspectives in its quest to explore relationships.

Tourism, recreation and leisure: the problems of definition

Writing over 20 years ago, Mathieson and Wall (1982: 7) commented that 'discussions of recreation and tourism are plagued by imprecise terminology'. Two decades on, it is apparent that whilst academic interest in the subjects of leisure, recreation and tourism has mushroomed, attainment of consensus on the meanings of these terms has, if anything, become an even more elusive goal.

Of the three, it is fair to say that the greatest uncertainties have focused upon the concept of leisure. Although this book is not explicitly concerned with leisure it is helpful, as we shall see shortly, to conceive of both recreation and tourism as areas of activity that are primarily (though not exclusively) located *within* leisure. Some discussion of the conceptual and definitional problems of the term 'leisure' is therefore warranted.

Traditional conceptions of leisure have been in terms of time – particularly those periods of the day that remain when responsibilities for work or domestic duties and the need for sleep and personal care have been discharged – and the activities that make use of that time. As many authors have indicated (see, for example, Clarke and Crichter, 1985) the origins of the concept of leisure as time lie in the emergence of capitalist–industrial economies in the eighteenth and nineteenth centuries. The factory system, in particular, imposed a rigorous demarcation between hours of work (in which the workers' time belonged to their employers) and remaining periods where usage of time was discretionary, whilst still being widely prescribed by the need to fulfil essential domestic or communal tasks. It was the rigidity of industrial patterns in regulating time – in contrast to the flexibility that characterised many pre-industrial systems – that encouraged the recognition of periods of leisure as an antithesis of work.

The conception of leisure as time defined in relation to work has been influential not only in demarcating periods in which leisure might be expected to take place, for example, evenings, weekends and designated holidays, but also in ascribing a range of associated values that have become firmly attached to the practices of leisure through the contrasts that they provide with work. Thus, leisure has traditionally been seen not only as embracing notions of reward for labour, but also freedom of choice, self-expression, self-determination and spontaneity (Rojek, 1993). Hence, leisure activity was often viewed as a catharsis, refreshing and recreating the individual for work (Smith and Godbey, 1991) yet also providing an arena of essentially positive and beneficial experiences that lay beyond the realms of working routines.

Additionally, the separation of work and leisure encouraged spatial distinctions through processes of development of spaces and facilities that were literally set aside for leisurely use. Provision of public parks, playgrounds, libraries and, later, sports fields – for example – are all symptomatic of the capacity of the modernist project to order time, space and its associated activities.

The validity of this outwardly simple view of leisure has, however, been progressively challenged. First, the construction of leisure in relation to paid employment has drawn criticism from feminists who complain – with justification – that because of the strong links between work and male identity, such a view perpetuates male hegemony and diminishes or disregards the unpaid work of women as homemakers or childcarers and the limited leisure opportunities that such roles permit (Deem, 1986; Wimbush and Talbot, 1988; Henderson, 1990). Readings of leisure when couched purely in terms of a relation to employment are, therefore, viewed as gender-specific and unrepresentative.

Moreover, second, the progression from an industrial to a post-industrial state has been associated with significant restructuring in the nature of employment that has made the meaning of distinctions between work and leisure – whether for men or women – much harder to sustain. Post-industrial change (see below) has been broadly associated with increasing flexibility in patterns of work, a flexibility that is revealed in the spatial organisation of activity, the size and composition of workforces, its functional skills and its temporal availability (Breedveld, 1996). The consequence is that traditional patterns of work are becoming blurred and amongst professional groups, at least, meaningful distinctions between work and leisure have become harder to draw (Harre, 1990; Rojek, 1995). The growing proportion of the population that is retired or which works on part-time contracts further undermines the value of simple relations between work and leisure.

Third, authors such as Rojek (1993) have questioned the basic notions that individuals may create genuine freedoms of action or expression through leisure, given the existence of complex and entrenched rules of self-governance and self-regulation that shape social actions. The progressive commodification of leisure within a commercial leisure 'industry' similarly challenges the scope for individual expression. This produces what Kelly (1991) defines as a 'false consciousness' in which people's leisure is actually dominated by a pervasive culture of commodification and control that actually erodes notions of freedom and choice, but where the restriction is not acknowledged or recognised by participants.

Fourth, the constitution of leisure as an essentially homogeneous area of playful, productive and relaxing activity has been redefined as a more complex picture of a diversity of leisure activities and objects has emerged, some of which conflict with notions of leisure as a source of social improvement. Thus, whilst many people have built a leisure lifestyle founded on a pattern of routine engagement with common leisure activities (Gerstl, 1991) and casual leisure (Stebbins, 1997), others pursue what has been recognised as 'serious' leisure (Stebbins, 1982). Here participants engage in the systematic pursuit of an interest in which they invest significant levels of resource (in time and money) and through which the acquisition of skills, knowledge and experience related to the activity replace (or at least match) the sense of fulfilment or achievement that is normally acquired through a career. (Roberts (1989) refers to this practice as 'connoisseur' leisure.) More recently, Rojek (1999) has argued that leisure is not even necessarily formed around pursuits that are unequivocally wholesome or good. There is, he suggests, an elective affinity between leisure and social deviance that needs to be acknowledged and understood.

For these reasons, definitions of leisure have tended to move away from traditional associations with non-work time and associated activities, and towards a construction that acknowledges the importance of personal attitudes and state of mind. This is not a new idea, but it is gaining wider acceptance as some of the deficiencies of other approaches become more evident. Pigram (1983: 1) notes that 'leisure means different things to different people' and that rather than being confined to so-called 'uncommitted time', it diffuses through a wide range of situations where it is contingent upon individual mood, intention and objectives. Put simply, leisure is what people make of it. Fundamentally, too, leisure is a *practice*, not merely a product to be consumed (Crouch, 1999), and conceived in this way it allows for recognition of important elements of human agency and the scope for reflexivity in the way in which individuals make their leisure – even in a world in which there are significant levels of commodification. In some situations, therefore, leisure practices become a site of resistance. Henderson (1990), for example, observes that leisure involvement for women may be a means of liberation from restrictive gender roles and, thus, a means for empowerment.

In comparison with the difficulties and nuances that infuse the definitions of leisure, recreation and tourism may be defined rather more easily although still, it must be noted, without quite reaching a consensus. 'Recreation' is most commonly connected with the idea of activity – with purposeful and constructive engagement with a pursuit or event. Pigram (1983: 3) considers recreation to be 'activity voluntarily undertaken, primarily for pleasure and satisfaction, during leisure time . . . [and where normally there is] . . . no obligation, compulsion or economic incentive'. It is implicit in most recreational activities that the participant derives some restorative benefit through participation – that they are *re*-created by the experience or its outcomes – and that recreation may deliver both intrinsic and extrinsic benefits. Accordingly, a very wide range of human interests and actions can be classified as recreations and, once again, personal perceptions and individual values may be crucial determinants of the status and meaning of a particular activity or event.

Tourism clearly involves recreation, but it also embraces other areas of behaviour and interaction. As Gilbert (1990) observes, the word 'tourism' draws complexity from its use as a single term to designate a variety of concepts. Implicitly tourism involves travel – the word 'tour' referring to a journey in which the traveller returns to the place of origin – and from this it may be deduced that the act of touring is a temporary condition that is dependent upon being away from home (see Murphy, 1985). However, the act of touring initiates particular sets of demands: for transport, travel services, accommodation, hospitality, entertainment and information, and these draw forth a complex pattern of supply of essential goods and services from what might be recognised as a tourism industry.

Several definitions have attempted to capture this holistic nature of tourism. For example, in 1979 the British Tourism Society (quoted in Gilbert, 1990) stated: 'tourism is deemed to include any activity concerned with the temporary, short-term movement of people to destinations outside the places where they normally live and work, and the activities during their stays at these destinations'. More recently, the World Tourism Organization (WTO) (quoted in Lickorish and Jenkins, 1997) stated: 'tourism comprises the activities of persons travelling to and staying in places outside their usual environment for not more than one consecutive year for leisure, business or other purposes'. In framing this relatively simple statement, the WTO is aiming to remove some perennial problem areas from the definition of tourism, in particular, uncertainties surrounding the inclusion (or otherwise) of day visitors or excursionists alongside staying visitors and associated questions of whether tourism should be defined by a minimum length of stay – as it has often been. Instead, this definition proposes a basic equation in which the tourist is a visitor (see Figure 1.1).

Figure 1.1 Tourism or recreation? Holidaymakers and day visitors pack the beaches at Brighton

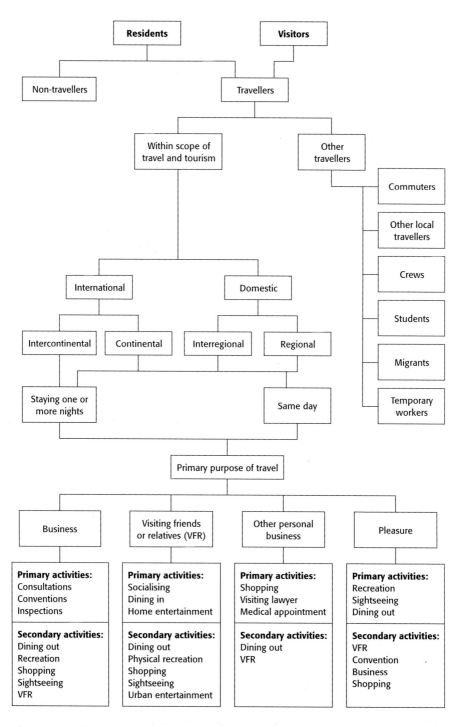

Figure 1.2 The structure of demand in travel and tourism
Source: Adapted from Chadwick (1987).

Clearly, therefore, tourism may be undertaken for a variety of purposes: pleasure, recreation, holidaymaking, visiting friends or relatives, business, education, health or religion, and may involve both domestic travellers (who move within their own national boundaries) as well as international tourists (who cross national boundaries). It follows, too, that 'tourists' do not constitute an undifferentiated mass, but rather there are distinctive types or categories of tourist who travel for different reasons and within differing organisational or social contexts. (For a fuller consideration of types of tourist, see *inter alia*: Cohen, 1972, 1974; Smith, 1977; Murphy, 1985; Shaw and Williams, 1994; Lickorish and Jenkins, 1997.) The complexity of the demand side of tourism is illustrated in Figure 1.2.

How are the three areas of leisure, recreation and tourism related? Figure 1.3 presents what is termed a Venn diagram in which the different areas are shown as intersecting spheres of activity. Conceived in this way, the diagram makes several fundamental points:

- By their nature, most areas of recreation and tourism are rightly located in the wider field of leisure. These are areas of experience that generally occur in what most people will identify as leisure time, and where they deliver many of the personal rewards and benefits that people ascribe to leisure activity.
- There do exist, however, areas of both tourism and recreation that extend outside the orbit of leisure and into areas of work. This is evident in the incidence of business tourism, but also in the realms of serious leisure where recreational interests take on some of the attributes of work in the quest for a 'professional' level of competence in the activity in question.

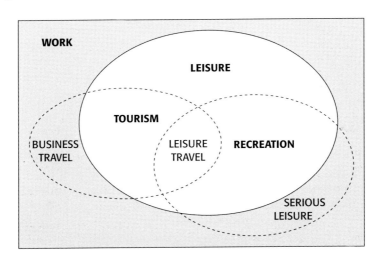

Figure 1.3 The relationship between leisure, recreation and tourism
Source: After Hall and Page (1999).

- There is a considerable degree of overlap between the spheres of recreation and tourism which is evident, at the most simple level, in the common coincidence of tourists and recreationalists in time and space.

This latter point, in particular, is central to the approach in this book. As mentioned in the introduction, academic perspectives on recreation and tourism have tended to perpetuate a sense that these activities are fundamentally different. Moore et al. (1995) suggest that some of the 'differences' are a product of the contrasting social and cultural histories within which leisure, recreation and tourism have arisen and consequently the different critical or theoretical positions from which they have been viewed. In practice, the differences are often more imagined than real. As McKercher (1996: 563) comments, 'both tourism and recreation [often] share the same resources, use the same facilities, compete for the same consumer dollars, exert similar impacts when the same activity is undertaken and produce common social and psychological outcomes for the participants'. Instead, McKercher (1996) proposes that tourism and recreation are more likely to exist on a continuum in which tourism that is not work-related may be seen as an extreme form of recreation, distinguished perhaps only by the degree of spatial separation between the leisure site and home, by the duration of stay, and by the individual psychology of the visitor in defining their perceived status as recreationalists or tourists.

Even within a continuum, however, clear differences may exist towards the extremes of the scale. Leiper (1979) has suggested seven criteria by which tourism may be distinguished from recreation:

1. The nature of withdrawal and return from the normal place of residence and its associated activities is more pronounced.
2. The duration of travel is often greater.
3. The frequency of tourist trips is less than for other recreational trips.
4. Tourism offers a wider range of social opportunities than are normally available through routine recreation and leisure.
5. The costs are generally higher.
6. The experience may be more exclusive.
7. The tourist trip is often perceived as discrete and therefore stays in the memory.

However, although these criteria offer relevant distinguishing dimensions, the argument that tourism and recreation are essentially related is not negated by Leiper's observations and as tourism has become a more routine phenomenon within orthodox leisure lives, the distinctive qualities have become less evident. As Moore et al. (1995) somewhat tersely observe, today there is nothing 'special' about tourism's 'specialness'.

Two further observations are relevant at this point. First, the use of the broken lines around each of the spheres of experience in Figure 1.3 is intentional – designed to reflect the uncertainties of definition that have been aired above and the fact that the boundaries are becoming progressively opaque as post-industrial change proceeds. The term 'de-differentiation' has been proposed to describe the blurring of conventional economic, social and behavioural boundaries (Lash and Urry, 1994) and will be used again in the discussion of some of the wider implications of this process later in this opening chapter. Second, it will be argued throughout the book that the overlap between recreation and tourism is becoming more pronounced with the passage of time. Historically, meaningful distinctions between tourism and recreation were arguably evident and generally recognised (although it will also be argued that the two were never completely separate). What has changed is the degree to which these two areas of leisure have converged – or become de-differentiated. This is evident within Figure 1.2 where, it will be noted, many of the activities that fill the time of the different categories of tourist at their destinations are recreational in character. At the level of the individual, therefore, distinctions between tourism and recreation are becoming irrelevant (Jansen-Verbeke and Dietvorst, 1987).

To explore more fully the nature of convergence – the 'common ground' between recreation and tourism – the next section of the chapter introduces five key areas in which the coincidence of recreation and tourism is evident: enablement; motivation; behaviour; space and time; and policy.

Tourism and recreation: finding the common ground

Enablement

Tourism and recreation are essentially enabled by common sets of factors that may be seen as encouraging – by facilitation – people's participation. Since these factors are generally well understood and are well covered in the literature, they may be outlined in summary form. They include: increases in leisure time; rising levels of affluence and personal mobility; changes in individual and communal expectations; and associated increases in the opportunity to pursue tourism and recreation through wider provision. (For general discussions of the individual and collective impacts of these variables see, *inter alia*: Gershuny and Jones, 1987; Hall and Page, 1999; Olszewska and Roberts, 1989; Patmore, 1983; Torkildsen, 1992; Williams, 1995.)

Increased leisure time

In general, the ability to participate in recreational and, especially, tourist activity is dependent upon the availability of periods of leisure time. Over the second half of the twentieth century leisure time increased noticeably: through a reduction in the commitment of time to employment; through the increased incidence of (early) retirement; through extended holiday entitlement; and a wider release from domestic duties. In the home, recreational time has been created through the virtual ubiquitous use of machines or automated systems (such as washing machines and dryers, deep-freeze units, automatic dishwashers and central heating systems) to perform routine chores. In the UK, hours spent at work have fallen with average hours worked by full-time employees (including overtime) dropping from around 45.5 hours per week in 1961 to 38.7 hours in 2000 (Central Statistical Office, 1990; ONS, 2001). Additionally, paid holiday entitlements – which are essential to the development of tourism – have risen. In the UK in 1961, 97 per cent of full-time manual employees enjoyed only the legal minimum of two weeks' paid holiday per year. By 1988, 99 per cent received at least four weeks' paid leave and of these, nearly a quarter received five weeks or more (Central Statistical Office, 1990).

Increased affluence

Rising levels of real wages and increased levels of disposable income within most westernised economies have increased the scope for purchasing recreational goods and services or tourist holidays and excursions. In the UK, the real household disposable income per head doubled between 1971 and 1997 (ONS, 1999), whilst data from the USA for one employment sector (manufacturing), show that wage levels increased by 642 per cent between 1955 and 1993. Personal savings – as indicated by levels of bank deposits within savings accounts – rose from US$30 billion in 1955 to US$579 billion in 1988 (Mitchell, 1998).

Evidence suggests that as lifestyles become more oriented towards leisure, so greater amounts of spending are being devoted to this area. In the UK, for example, average weekly household expenditure on leisure goods and services rose from £28.20 in 1970 to £54.40 in 2000 (at 2000 values), although in the top 10 per cent of households in terms of income, average weekly expenditure on leisure exceeds £160 (ONS, 2000). These values (which exclude holidays) constitute an average of 15 per cent of all household expenditures, but in the wealthiest households the share rises to over 20 per cent.

In the USA both the level of spending and the rate of recent change have been more impressive still. The total value of recreational expenditures rose from US$284 billion in 1990 to US$494 billion in 1998. In addition, a further US$426 billion was spent on domestic vacation travel in the latter year and some US$76 billion on foreign travel – a rise of nearly 58 per cent since 1990 (US Dept. of Commerce, 2000).

Increased personal mobility

Rising levels of affluence have also had a direct bearing upon changes in levels of personal mobility, especially through widening ownership of cars. The growth of globalised mass production methods and the development of a sizeable market in second-hand vehicles have seen car ownership extend across the social spectrum. In the UK in 1961, almost 70 per cent of households were without a car whereas by 1997, 70 per cent possessed at least one car and a quarter owned two or more vehicles. There are now more than 20 million cars on British roads (ONS, 1999) and comparable (or greater) rates of ownership are found elsewhere. In the USA, there are more than 150 million cars (Mitchell, 1998) and in France and Germany, 25 million and 41 million cars respectively (UNDESA, 2000).

The effects of car ownership upon recreational patterns are significant. Enhanced mobility extends the geographical range over which people may travel for recreation and brings more resource areas into use. In the UK, around 75 per cent of all domestic tourist trips are made by car, as are over 60 per cent of recreational journeys. As Hall and Page (1999) observe, the concept of 'proximity' to resources is flexible, varying with the speed of travel, and as travel has accelerated so proximity has been redefined. Thus domestic tourism has benefited from the spread of motorways, whilst international travel has become easier through the development of accelerated rail services (for example, across much of the European Union) and especially from the spread of low-cost, charter-based air services. Today, a tourist commencing a journey in London can reach the west coast of the USA by air in the same time that it would have taken to reach the far west of Cornwall by road or by train 50 years ago.

Changing expectations of leisure

Mobility and affluence directly enable participation, but the desires to engage are fostered by more fundamental changes in society whereby recreation and tourism have become progressively more central in the construction of everyday lives. Changes in the workplace that surround not only the reduction of hours but also processes of deskilling or the wider incidence of part-time work and early retirement have, it has been argued, reduced the role of work as the context in which social behaviours are shaped and regulated. Instead, in the post-industrial world of the late twentieth century, leisure became a more influential nexus in structuring time, creating interests and shaping and maintaining social contacts (Haworth, 1986; Reid and Mannell, 1994). In this way, tourism and recreation have come to be viewed as definable social needs that satisfy a range of motives that are embedded in contemporary lifestyles.

Wider provision of opportunity

In most instances, demand for recreation and tourism will only become evident if there is a supply of suitable facilities, so part of the explanation for the levels of contemporary activity lies in the widening base of provision. In the field of urban recreation, for example, traditional resources (such as town parks, recreation grounds, playing fields, swimming baths, public libraries and a range of commercial attractions such as cinemas, theatres, restaurants and bars) have been supplemented by newer forms of provision. In the UK, this has included a significant expansion in publicly provided indoor sports centres (often with swimming pools), as well as developments that reflect the post-industrial transformations of economy and society. These include the growth of leisure shopping (especially in malls and out-of-town retail parks) and the enjoyment of new forms of heritage-based recreation based around the remembrance and celebration of former sites of industrial production and transportation. The conversion of industrial buildings into museums, restaurants, galleries or shopping arcades; or the wholesale redevelopment of former dockland as new zones of residence, amenity and recreation are all illustrative of this trend and draw both tourists and local recreational visitors with seemingly equal capacity.

Motivation

A second area in which there is significant coincidence between recreation and tourism is motivation. The question of *why* people participate in recreation and, particularly, tourism, has engaged leisure scholars for several decades and drawn forth a diverse range of conceptualisations and explanation. Many of these discussions implicitly develop the notion that recreation and tourism – as a part of leisure – aim to meet desires for temporary escape from work-based or domestic routines and provide essential contrasts through which individuals may develop self-expression, acquire new skills and enjoy new experiences. This view is articulated, for example, in Iso-Ahola's (1982) well-known model of tourist motivation in which activity is seen as a product of a push–pull relationship – the participants aiming for escape from everyday personal and interpersonal environments whilst also seeking intrinsic rewards from experience gained at the places they visit. Escape and contrast with the routine are also implicit in Graburn's (1983) explanation of what he terms the tourist's behavioural 'inversions'. Here Graburn argues that in leisure situations – especially holidays – people commonly adopt behavioural patterns that contrast with their norm – partly for basic hedonistic reasons, but partly also to distinguish the tourist event. Thus, for example, people may sleep in, overindulge with food and drink, spend abnormal amounts of money on entertainment, adopt unusual dress codes and visit sites or places that they do not normally frequent.

Embedded within these wider motives are a diverse range of more specific stimuli to participation, including a quest for: relaxation and exercise; pleasure and enjoyment; social contact; mental stimulation; improved health; personal advancement; social recognition and enhanced self-esteem (see Argyle, 1996; Hall and Page, 1999: 26–30; Krippendorf, 1987; Pearce, 1993). However, what is especially relevant here is that the identification of motives for parti-

Table 1.1 Comparison of motives for participation in leisure/recreation and tourism

Leisure/recreation	Tourism
Escape from routine	Escape from mundane environments
Health and fitness	Relaxation, recuperation or health
Mental stimulation/enjoyment	Opportunity for play or entertainment
Promote family activity	Strengthen family bonds
Gain positions of leadership	Acquire prestige or social enhancement
Make new friends/enjoy company	Social interaction/meet people
Acquire new skills and abilities	Education
Enhance self-esteem/status	Fulfil wishes or ambitions
Meet challenges	

Sources: Adapted from Kabanoff (1982) and Ryan (1991).

cipation in tourism on the one hand, and in leisure activity and recreation on the other, is essentially similar. Table 1.1 lists the basic motivations proposed by Kabanoff (1982) in respect of leisure and recreation and by Ryan (1991) in relation to tourism. The similarities are rather more apparent than the contrasts.

The sources cited above generally present what we may term 'established' (although still valid) views of recreational and tourist motives. More recently, other explanatory ideas have gained credence and one – the maintenance of identity – is worth further discussion. Miller *et al.* (1998: 19) comment that identity has become 'one of the key words of the 1990s'. It has become essential to the social recognition of difference, to the cultural constructions of 'others' and, of particular importance, to the construction of the self, i.e. our own image of who we are. This latter process is generally termed 'self-affirmation'.

Haggard and Williams (1992) explain that self-affirmation is the process by which we construct an identity image (or set of images) for ourselves that we then strive to affirm both to others and, in the process, to ourselves. Self-affirmation behaviours tend, therefore, to become quite pervasive, being manifest in aspects such as our appearance, our use of clothing and the manner in which we interact with others. However, because freely performed behaviours influence one's self-perception more so than constrained behaviours, leisure activity becomes a particularly potent area for self-affirmation (Wearing and Wearing, 1992). As a partly subconscious process, many participants would fail to acknowledge the affirmation of identity as a motive for participation in recreation and tourism. Yet it is evident from both the recreational and tourist choices that people make, and the manner in which experience is recounted to others, that self-affirmation is a key motive. This is especially true in tourism where many people routinely seek to assert identity and acquire social status by relating their tourist experiences to friends, relatives or work associates, but it is present in many areas of recreation too – for example, in the clubbing activities of young adults in urban communities or amongst supporters of professional team sport.

Behaviour

Much of the main content of this book is explicitly and implicitly concerned with relationships in behavioural patterns of tourists and recreationalists – especially through their coincidence in space and time (see below).

However, less obviously, but no less importantly, there are marked behavioural similarities encapsulated in the nature of the experience that surrounds recreational and tourist events. As a concept, the nature of the experience has enjoyed only sporadic attention, despite the fact that it was articulated as long ago as the 1960s by the pioneer American recreational scholar Marion Clawson (1963). Yet it is fundamental to understanding behavioural patterns and, especially, the values that participants derive from recreation or tourism. In particular, the concept emphasises that the experience is multidimensional (Lee et al., 1994) and is shaped by a range of activities that *precede* participation; by a number of experiential inputs *during* participation; and important phases of recall *after* the event and which help to shape future patterns of behaviour.

Figure 1.4 sets out a recent reworking of the concept in the case of tourism (Williams, 1998). It shows how the trip is prefigured by a planning phase and then a journey (which is often an integral part of the experience for tourists) and followed by important processes of recall and memory which directly inform decisions about future tourist trips. In general we may expect positive experience to encourage repeat visits, or visits that build upon initial experience, whereas negative experience will prompt a reappraisal and, in all likelihood, alternative choices being made in the future. At the heart of the model lies the primary element, the experience of the tourist at the destination which, as the model shows, generally combines a range of elements that define the essential nature of the trip.

Interestingly, however, very similar models have been developed – quite independently – in the context of recreation. Figure 1.5 depicts a model of the recreational experience developed by Kay and Moxham (1996) in the specific context of recreational walking. Whilst there are differences of detail – for example, the recreational model emphasises environmental and social inputs to the central area of the activity itself – the important phases of planning and anticipation, together with recall and reminiscence, are replicated in the recreational context and may be visualised as working in broadly comparable ways to the tourist experience.

These models suggest, therefore, that whilst participants may often attach greater significance to their activity as tourists – for the reasons given earlier in relation to costs, duration, separation from the home and the 'special' status that is often associated with tourist excursions – the behavioural processes that contextualise tourist and recreational activity are fundamentally similar.

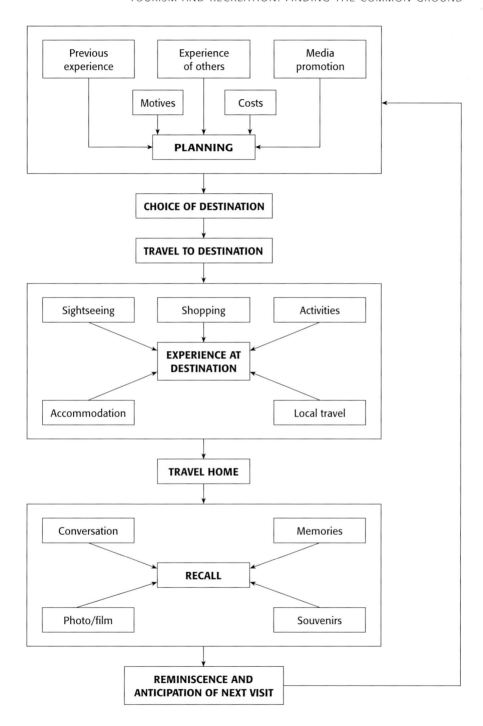

Figure 1.4 The nature of the tourist experience
Source: After Williams (1998).

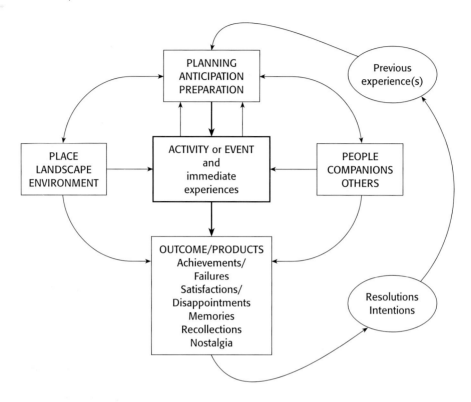

Figure 1.5 The nature of the recreational experience
Source: After Kay and Moxham (1996).

Space and time

In earlier phases of development up to perhaps the beginning of the twentieth century, tourism and recreation were differentiated quite effectively through space and time. Prior to the era of large-scale car ownership and affordable motoring, recreation was often a local phenomenon since travel to more distant places was less easy and seldom quick. Tourism centred upon travel, but the slow relative speed of most journeys, especially in comparison with those attained today, meant that tourism required time away from home. This pattern contributed to a traditional differentiation in which recreation was likely to be local and regular, whereas tourism would be distant and occasional.

However, the acceleration of travel: through the replacement of steam trains by diesel and electric services on the railways; through the development of interregional and international motorway links; and through the substitution of jet planes for piston-driven designs, has combined to produce significant time–space compression. The consequence of this compression has been to produce a greater spatial coincidence between recreational day visitors and staying tourists. This is due partly to the erosion of the local–distant sense of contrast, as well as the closer alliance of recreation and tourism that has emerged since the 1960s and which often promotes the same practices and attractions to tourists and recreationalists alike. This has been particularly evident within domestic markets and, to a lesser degree, in shorter-range international visits – such as between Britain and the near continent of Europe. Space is still a differentiating barrier, but the thresholds at which that differentiation occurs are now greatly extended in comparison with earlier patterns and are more likely to be set in the international rather than the domestic context. Thus, in domestic markets in the UK, for example, much more complex spatial coincidences of recreation and tourism have emerged because key holiday tourist regions (such as the south-west of England) can be reached easily by car in under three hours from major generating areas (such as London, South Wales or the West Midlands). Tourist areas have become, therefore, areas for recreational day visits too. (Some of the consequences of this changed level of accessibility for British coastal resorts are explored more fully in Chapter 2.)

Temporal distinctions have also been eroded. Similarities in the temporal patterns of many forms of recreation and tourism owe much to the fact that both practices are closely linked to the concept of leisure time, which remains quite firmly associated with core periods: the evening, the weekend, the school holiday, and the months of high summer. Additionally, however, as recreation and tourism become more intimately woven into the fabric of daily lives through greater amounts of leisure time and a greater frequency with which tourist excursions are made, so the notion of clear compartmentalisation of activity into defined moments becomes less meaningful.

Part of the explanation for change in this area lies in the emergence of a post-industrial economy and society. This key area of change has been responsible for a fundamental reworking of many of the traditional relationships between work and leisure and between people and place. It has therefore been responsible for the creation of many new opportunities for recreation and tourism, in both time and place.

The theme of a post-industrial transformation will be revisited at several points in subsequent discussions, but an initial outline of the process may be helpful at this stage. The basis of change lies primarily in the emergence of new patterns of capital accumulation centred around a shift in the economic rationale of urban places from production to consumption, from a manufacturing to a service economy, and with associated moves from mass to individualised forms of behaviour. Patterns of production and consumption are characterised by flexibility in which, for example, the increased use of short-term contracts of employment and part-time working allows employers to vary the size and capabilities of the workforce to suit short-term changes in economic conditions. So not only have many sites of former industrial production been abandoned, so too have many of their working practices.

This process has initiated an important reworking of the physical space of cities, creating new spatial patterns of opportunity. This has occurred especially through conversion of former sites of industrial production and distribution to new centres of consumption or residence – for example, within former dockland areas. Additionally, though, the post-industrial transformation has also been associated with influential changes in social and cultural formation and reproduction. In particular, work has been widely replaced by areas such as leisure as an organising feature of daily life and a basis to the formation of personal and collective identities within consumer-based societies (Featherstone, 1991). Hence, the 1980s (in particular) witnessed the rapid emergence of new urban elites – people whose positions owed little to the traditional pathways through which upwardly mobile groups were previously formed (Rose, 1984), but whose almost incidental strategic centrality within new service industries (such as telecommunications) helped to form new patterns of urban wealth and permitted the development of an influential consumer culture. Within this new consumer culture, lifestyles were often positioned around the conspicuous consumption of leisure, much of which was located in new times and new spaces. Moreover, as Breedveld (1996) argues, the increased de-differentiation of work and leisure that is one of the hallmarks of post-industrial change places a clear premium on the context in which actions occur and the individual's interpretations of those actions. In this sense, the times and places of recreational and tourist practices are thus less clearly demarcated than ever.

Policy

Post-industrial change links clearly into the final area of emerging common ground between recreation and tourism – the area of policy. Due to its essentially reflective nature, the policy responses to the types of change described above have inevitably taken up a complementary agenda. This has been evident across the range of urban and rural environments in which policy that involves recreation and tourism has been developed.

In the urban context, for example, Mommaas and van der Poel (1989) – in reviewing recent experiences in the Netherlands – show how urban policy has developed new perspectives on the city as a commodity to be marketed. Typically this approach has revealed a new role for powerful and influential private–public partnerships in bringing eye-catching development schemes to fruition, but the cumulative effect has seen extensive development of new sites of production and consumption: in converted docklands (Figure 1.6); in restored mills and factories; in new shopping malls and entertainment centres; and in state-of-the-art sports stadia. In many cases these developments have built upon local patterns of recreational demand, but in the process, leisure has become more than just an instrument of economic development. Rather, it becomes a key component in shaping and defining marketable images of the city that aim to draw population and inward investment, as well as attracting

Figure 1.6 Converted dockland in Gloucester provides a new set of leisure sites based upon heritage museums, restaurants, bars and shopping arcades

tourists from outside. Thus, there exist genuine links in which responses to new patterns of local recreational activity have shaped spaces that appeal with equal force to tourists. These issues are explored more fully in Chapters 4 and 6.

Similarly, the synergy between recreation and tourism has also become apparent in many areas of rural policy. Rural areas across Britain, much of Europe and, to a lesser degree, North America have witnessed important changes in their economic, social and cultural functions since the end of the 1960s. There is now less emphasis upon traditional areas of production and a much greater concern for developing the amenity value of the countryside as a place in which to live and in which to enjoy recreation and tourism. This has promoted the development of new policy approaches that aim to integrate recreation and tourism into the rural economy alongside older areas of production in agriculture and forestry. As the numbers who visit the countryside for leisure have swelled, however, so the need for policy that aims to conserve and protect fragile environments has also gained significance.

These concerns are mirrored in several policy areas affecting both rural development and conservation and in which recreation and tourism are discussed in ways that acknowledge their close relationship. As an example, Chapter 6 will explore some of the policy issues that surround the development of public access to rural land, whilst Chapter 5 illustrates in greater detail how recreation and tourism are being integrated into the new patterns of the 'post-productive' countryside.

Recreation and tourism: convergence and de-differentiation

In light of the extensive range of common ground that the previous discussion has identified, the natural question to ask concerns the extent to which tourism and recreation remain as differentiated activities. Conventional understandings of tourism in general (and foreign tourism in particular), argue that its development has been driven by a quest on the part of the tourist to experience difference and, in some readings (especially MacCannell, 1989, 1992), by a desire for *authentic* experience. As Rojek and Urry (1997b) explain, tourism derives (or derived) at least part of its meaning through contrasts with other forms of leisure or more general facets of life. Thus, not only could tourism be differentiated from areas of work or domestic commitment, but also from leisure pursuits such as day-tripping or more esoteric activities such as exploration. However, as increasing proportions of the population experience both domestic and foreign forms of travel as integral parts of a cosmopolitan lifestyle (see Featherstone, 1991), so that sense of difference has become blurred. This has occurred in two ways: first through the importation of tourist experience into day-to-day leisure and recreation, and second, via relatively simple processes of familiarisation that reduce notions of difference and the special status of tourism in most peoples' leisure lifestyles.

The influence of foreign travel upon day-to-day leisure and recreation has been especially marked, even though it has often been an insidious process. The mechanisms are complex, but the net effect is often a seamless integration – into both our working and leisurely lives – of fashions, tastes, knowledge and practice acquired through the medium of travel, especially abroad. This may have a number of effects, including:

- Introduction of new influences over recreational tastes and preferences, leading to new patterns of local demand. This is evident in areas of conspicuous consumption such as eating habits, where not only has there been a dramatic increase in countries such as Britain and the USA in the incidence of restaurant use, but also in the diversity of foreign food restaurants that are available in most cities (Pillsbury, 1990). Some areas of recreational activity have also been affected by tastes derived through foreign travel. The provision of dry ski slopes in the UK, for example, is almost entirely a response to demands fuelled by the experience of Alpine and North American winter sport holidays.

- Greater levels of resistance towards established but traditional forms of leisure and recreation. (The decline of the cold-water seaside resorts of northern Europe is an examplar here – see Chapter 2.)

- Importation of passive consumption of travel experiences into domestic leisure patterns. This is evident especially via leisurely use of the media. Many hundreds of hours of terrestrial and satellite broadcasts are devoted to programmes that present tourist places to armchair consumers, whilst magazines, newspapers (especially the broadsheets) and tourist guidebooks are similarly comprehensive in their coverage of foreign places, their cultures, societies, histories, natural histories and their travel opportunities. As Urry (1990b) emphasises, the use of such media in anticipation of foreign visits is an integral part of the overall tourist experience and may therefore constitute a significant use of leisure time prior to a visit. Similarly, scenes captured in photographs, postcards or on video enable experiences to be endlessly reproduced and recaptured – once again, in routine leisure time (see Figure 1.4).

Furthermore, a surprising range of daily routines (both leisurely and non-leisurely) are grounded in contexts that depend upon a tourist dimension as a key element in their construction and in which there is a presumption that a significant proportion of the population possesses the knowledge or experience to 'read' and react to the signs and stimuli that are presented. The contemporary shopping mall offers an increasingly familiar example, as a leisure site in which collages of signs and references are drawn from across the globe in what Urry (1995) describes as an extreme form of global miniaturisation. Lury (1997) provides another example in the promotion of household goods by UK retailers such as Boots (chemists) and Habitat (furnishings and household goods). Here, two practices are evident: first, the naming of products with titles that conjure a sense of exotic places; and second, the use of visual images in packaging or product catalogues that explicitly and implicitly offer 'authentic' representations of other cultures or places. The veracity of these claims is largely immaterial – what is important is that such representations succeed because people are able to relate to the images presented, and whilst such awareness comes through a multiplicity of channels, the direct experience of tourism itself is one of the more significant.

The sense of convergence – the merging of tourism and recreation within the overall framework of leisure – can be construed as one element in a much broader process of post-industrial change that has seen an erosion of conventional distinctions in all kinds of social, cultural and even economic spheres. De-differentiation focuses attention on a plethora of contexts in which formerly clear distinctions (such as work/leisure, home/abroad, elite/popular, public/private) are eroded as cultural, social, economic and geographic spheres increasingly overlap (Lash and Urry, 1994; Urry, 1995; Rojek and Urry, 1997b). Travel grew partly because the desire to experience difference could only be and issues that are associated with recreational and tourist use of places are addressed through policy.

Questions

1. Explain why the concept of leisure is complex.
2. To what extent do you believe Leiper's (1979) criteria for differentiating tourism from recreation remain valid?
3. What are the key factors that have promoted the growth of leisure and tourism since 1970?
4. In what ways are recent changes in recreation and tourism a reflection of a post-industrial transformation in the wider economy, society and environment?
5. What do you understand by the term 'de-differentiation'? How far do you think it is valid to consider contemporary recreation and tourism to be de-differentiated areas of activity?

Further reading

For an excellent discussion of the evolution and place of leisure in modern society, see Clarke, J. and Crichter, C. (1985) *The Devil Makes Work: Leisure in capitalist Britain*, Basingstoke: Macmillan. A more complex argument on essentially the same theme is provided by Rojek, C. (1995) *Decentring Leisure: Rethinking leisure theory*, London: Sage.

A convenient summary of the basic nature of recreation and tourism is provided in Hall, C.M. and Page, S.J. (1999) *The Geography of Tourism and Recreation: Environment, place and space*, London: Routledge.

A stimulating book on the subject of tourist motivations and behaviour is provided in Krippendorf, J. (1987) *The Holiday Makers*, Oxford: Butterworth Heinemann.

Rojek, C. and Urry, J. (eds) (1997) *Touring Cultures: Transformations of travel and theory*, London: Routledge offers an interesting collection of essays on the place of recreation and tourism in the postmodern world.

The concept of de-differentiation is explained in both Lash, S. and Urry, J. (1994) *Economies of Signs and Spaces*, London: Sage and Urry, J. (1995) *Consuming Places*, London: Routledge.

Seaside resorts as centres of tourism and recreation

CHAPTER 2

Introduction

In his excellent review of the historical geography of recreation and tourism, John Towner (1996: 167) correctly observes that 'the lure of the sea and the consequent creation of seaside resorts have been the most impressive manifestations of the power of leisure to create new landscapes, to shape new patterns of activity, and to create new social and economic relationships'. For many people living in modern, industrial communities, the coastline and especially its seaside resorts constitute distinctive and special leisure places. This is a product of several factors – the unique attraction of the physical environment of the coast itself with its particular sounds, smells, sensations and views; the emotional response that the sight of the sea elicits from visitors; and the powerful –albeit traditional – sense of association between the seaside and holidays, those special events in the annual leisure calendar. It is the link between resorts and holidaymaking, in particular, that encourages us to think of resorts as tourist places, but as this chapter will attempt to show, such associations are too simplistic. The seaside resort is not just the product of a set of practices that we would now label as 'tourism', but a more complex synthesis of the special dimensions of tourism with the more routine and familiar forms of recreation. This synthesis, it is contended, has been true from the earliest stages of resort formation.

To elaborate this view, the discussion will pursue three related strands. First, the processes that shaped coastal resort development in Britain, parts of Europe and the USA are summarised to provide an essential context for the main argument. Then, and with particular reference to Britain in the nineteenth century, we shall explore the manner in which the contemporary recreational tastes and practices helped to shape the form of tourism that developed in resorts. Finally, the chapter concludes by considering how the changing roles of resorts in the late twentieth century have altered still further the balance between tourism and recreation at the coast.

function as leisure places, they were not replaced by seaside resorts. Borsay (1989) has shown how, during the latter part of the seventeenth century, the general rise in national wealth that accompanied the spread of a free-market capitalist economy and greater levels of political stability, fostered a cultural renaissance within the urban environment. This underpinned the emergence of a new leisured lifestyle amongst the urban elite that was sufficiently well established by 1750 to be quite capable of sustaining both the established spas and the new centres of leisure at the sea. Towner (1996) illustrates, for example, how the decline in the fortunes of Bath as a tourist spa was a protracted process, extending over more than a century (from about 1790), whilst in some parts of Britain spas were still being actively developed as late as the middle of the nineteenth century – at places such as Buxton and Harrogate in England, and Llandrindod Wells in Wales.

The spas did, however, exert a powerful influence over the early patterns of physical development at sea resorts and the various practices (both medical and leisurely) that took place. Although many of the sea-water treatments were physically unpleasant, uncomfortable and most definitely not considered as leisure, the new resorts quickly acquired the air of leisure that had already evolved at fashionable spas. This was mirrored in the construction of public spaces for promenading as well as the indoor facilities that polite society expected as a basis for its leisure – the theatres and assembly rooms for dances, music, gaming and other forms of entertainment, the libraries, meeting rooms and the coffee houses. The practice of visiting for a season was also transferred from the spas to the resorts, establishing not only a supplementary need for accommodation and other visitor services, but also a temporal pattern of patronage. All of these elements have persisted, although often in modified forms, to the present.

The organisation and provision of such facilities required both enterprise and, ideally, patronage. One of the reasons why resorts such as Brighton developed so strongly in the latter part of the eighteenth century was that they enjoyed royal patronage. Brighton became a favoured resort of the Prince Regent (later George IV) and therefore drew the elite society that followed members of the court (Figure 2.1). Similarly, Weymouth benefited from a visit by the King (George III) in 1784 whilst lesser princes and princesses patronised fledgling resorts at Sidmouth, Worthing and Southend (Pimlott, 1947).

Less conspicuous, but much more important, were the local businessmen and entrepreneurs whose investment underpinned the early provision of facilities to meet the needs of their wealthy visitors. In later phases of resort development, much of the capital and enterprise came from outside the resort, but initially, growth was dependent upon local investment and business acumen.

By 1800, a basic geography of resorts had begun to be established in Britain and was emerging as a parallel, though slightly later, development along the northern shores of Europe. In England, the particular appeal of equable climate combined with relative proximity to the major market for wealthy visitors

Figure 2.1 The Brighton Pavilion – built for the Prince Regent (later King George IV) – illustrates the significance of royal patronage in the early phases of English resort development

(London) had encouraged resort development along the southern coasts of Kent and Sussex, although smaller and less fashionable resorts that relied upon the patronage of local gentry had also been established further afield at places such as Exmouth (Devon) and Scarborough (Yorkshire). The taste for sea bathing had quite quickly crossed the English Channel to northern France and what is now Belgium, with resorts being developed at Ostend, Boulogne and Dieppe before 1800. In the USA, British colonial influence ensured that early phases of resort development followed familiar patterns, with small resorts developing on shorelines that were accessible to major urban centres, such as Nahant (north of Boston), Long Branch (south of New York) and Cape May (south of Philadelphia) (Towner, 1996). The following case study examines the early development of sea bathing resorts in the German states on the North Sea and Baltic coasts and draws both parallels and contrasts with the pattern observed in Britain.

Case study: The early development of sea bathing resorts in Germany

Although the German states had a tradition in the use of spas that was more extensive and well established than was the case in Britain, the habit of sea bathing at coastal resorts was taken up later and more selectively. Soane (1993) attributes this to the more conservative outlook of the German middle classes, who were less willing to indulge the pleasure principle that was shaping British and French resort development by the end of the eighteenth century, and who – in the quest for health – were generally content to remain patrons of inland spas. Thus, it is not until 1794 that development of the first German sea bathing resort at Doberan is begun.

As was true in Britain, the stimulus to the development of German resorts came first from members of the medical fraternity, at least one of whom – Dr Samuel Gottleib Vogel – had spent time in England and been influenced by resort development there (Corbin, 1995). The German coast offered two contrasting areas for development – the North Sea and the Baltic coasts – and the merits of each in respect of their potential curative qualities were widely debated. The advocates of the Baltic coast were generally more persuasive and as a result, the first German resort was commenced at Doberan (near Rostock) in 1794 (Figure 2.2).

British resorts generally developed in an incremental fashion, reflecting the sporadic manner in which demand emerged, as well as the limitations imposed by the availability of local capital, suitable land, and enterprise. In the German states (as in the Netherlands and Belgium) the process was more organised, with resort development typically being guided by companies of merchants, civil servants or physicians. However, as in Britain, royal patronage was often an essential component too and although the rulers of the different German states were not always enthusiasts for bathing, they often provided both encouragement and material support. The King of Prussia took a personal interest in the development of Colberg (1802) (Towner, 1996) and at Swinemunde (1822) appointed the directors, granted subsidies and contributed to the embellishment of the bathing club (Corbin, 1995).

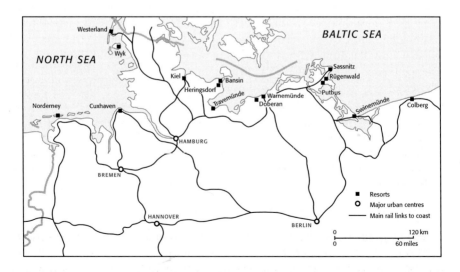

Figure 2.2 Major seaside resorts in Germany, circa 1900

Without exception, resort development in the German states was on a smaller scale than was generally found elsewhere, and even by the end of the nineteenth century, numbers of visitors were surprisingly low. Then the larger resorts such as Doberan and Norderney (on the East Frisian islands in the North Sea) attracted fewer than 50,000 visitors a year (Towner, 1996) compared with levels in England that saw as many as 2 million visitors at Blackpool (Walton, 1983) and figures approaching 1 million at popular southern resorts such as Southend (Walvin, 1978). This lower level of patronage of German resorts owed something to the constraints posed by a transport system that was less well developed than in Britain. Even at the start of the twentieth century, some German coastal resorts were still only accessible by horse and carriage, or by steamboat in the case of resorts on the offshore islands. But it is clear that cultural factors were also at work.

As a consequence of the smaller scale of the German resorts, their catchments were similarly restricted to local or subregional markets during their formative and developmental stages. Resorts such as Swinemunde relied heavily upon Berlin as a source for its visitors, and at several resorts people from outside would encounter significant levels of use by local residents, especially at weekends. At Doberan, for example, citizens from nearby Rostock had been accustomed to using the seashore for bathing and simple kinds of picnic before the resort was formed, and these practices appear to have continued after other visitors began to use the new resort. The principal adjustments that became necessary reflected a segregation of users on the beach according to gender and status, to ensure that the expected proprieties were duly observed (Corbin, 1995).

However, although levels of patronage were much lower, two points of similarity with the British pattern of development did emerge. First, although there was some local usage by ordinary people, the social differentiation of resorts that became evident in Britain (as well as in France, the Netherlands, Belgium and the USA), also emerged in German resorts. Hence, on the Baltic coast for example, Heringsdorf developed as an exclusive, expensive resort, whilst neighbouring Bansin served a popular, cheaper market (Towner, 1996). Second, the pattern of development within resorts revealed a similar progression from fashionable health spa to place of leisure.

Contemporary accounts of a number of German resorts dating from the 1820s show how the bathing establishments were soon augmented with hotels, promenades, piers, gardens, theatres, libraries, reading rooms, concert halls and meeting rooms. These venues provided the arenas for a well-developed pattern of social activity that included plays, concerts, dances, tea parties, reading groups and games of billiards, and which reflected very precisely the leisurely lifestyle of the patrons of these early resorts. These links between popular leisure patterns and the early practice of tourism are important in defining the links between recreation and tourism, and are explored more fully later in this chapter.

The early resorts – whether in Britain, the USA or along the northern shores of Europe – were characteristically small in scale and exclusive. Exclusivity was a product of several factors: the absence of time and the means to travel to relatively distant places (as well as the high costs of travel before 1815) kept most ordinary people away from the resorts, but their absence was also a consequence of the fact that, with some local exceptions, the populace at large had yet to acquire the taste for seaside leisure. After 1815, however, this pattern changed radically as seaside resorts developed initially as leisure spaces for an important, new, middle-class market, to be followed by the 1850s by the emergence of a mass market of working people. Consequently, the tone and character of these places shifted from being places of generally refined celebration of polite society, genteel recreations and picturesque seascapes to what Shields (1990) has described as zones of the carnivalesque – places of hedonism devoted to fun and entertainment. Central to this transformation were the revolutions in transport and the processes of social emancipation that proceeded over the course of the nineteenth century.

The development of popular resorts that served mass markets depended to a considerable degree upon new systems of transport that were capable of moving large numbers of people quickly and at low cost. The first of the new forms of transport to make an impact were the steamships which appeared on major waterways such as the Thames, the Clyde and the Forth from about 1815. In Scotland, small local resorts had formed at places such as Portobello (Edinburgh) prior to 1800, but the effect of new steamboat services, especially on the Clyde, was to stimulate significant expansion of resorts at places such as Greenock, Helensburg and Largs (Durie, 1994).

Opinions on the real impact of the railways on resort development vary. Towner (1996) seeks generally to play down the role of railways, arguing that in isolation, any transport link is insufficient to create new patterns of tourism without associated changes in social organisation and that in many cases, the railways simply reinforced existing lines of communication, modifying patterns rather than creating them. Urry (1990a) agrees that it is easy to overemphasise the impact of the railway, noting that it was not until after Gladstone's Railway Act of 1844 that cheaper fares became a reality and not until the end of the nineteenth century that active promotion of leisure travel and the development of extensive holiday traffic became routine parts of railway operations.

However, the capacity of the railways to move large numbers of visitors to resorts quickly and at relatively low cost proved a significant advance over the much more limited capacities of stagecoaches, and the railways increased the scale of seaside visiting, widened the clientele of resorts and promoted significantly the scope for day trips to the seaside. Rail excursions to Brighton at Easter 1844 delivered over 15,000 trippers to the resort, part of an estimated quarter of a million visitors who travelled to Brighton during that year. In north-west England, excursionist traffic was even more substantial. During Whit week in 1848, excursion trains took an estimated 116,000 trippers from Manchester to the coastal resorts of Lancashire (Walvin, 1978).

Walton (1979) argues that railways not only reinvigorated established resorts (such as Brighton) but also gave extra impetus to newer resorts that were growing. The value of a rail link is nicely illustrated in Devon where the contemporary resorts of Torquay and Ilfracombe developed very differently as a consequence of their transport links. Torquay was connected to the South Devon Railway in 1848 and flourished as a consequence, whereas on the north coast the growth of Ilfracombe was significantly constrained by the absence of a rail link before 1874 and its continued reliance on steamer services from Bristol and South Wales (May, 1983).

At first, the patrons of the new rail-based visits to the seaside were drawn from the emerging Victorian middle classes. Soane (1993) observes how the expansion of the international economy in the first half of the nineteenth century created new professional and service classes who enjoyed significant increases in wealth and associated leisure time and who quickly shaped consumption-oriented lifestyles in which family holidays to seaside resorts became an established feature. Out of such demand, fashionable resorts such as Eastbourne, Bournemouth and Torquay grew from about 1850.

Working-class holidays emerged more slowly and since such demand tended to be more constrained geographically and revealed stronger attachment to local practices, there was significantly greater variation in its spatial impact. Walton's (1983) study of working-class seaside leisure shows how the popular adoption of day excursions and (eventually) staying holidays at resorts emerged most strongly in the cotton towns of Lancashire. High levels of family income, local traditions in self-help and savings schemes and an established pattern of holidays centred around the custom of observing wakes weeks (Poole, 1983), permitted these groups of workers to be amongst the first to translate occasional day excursions to the seaside into longer visits. In contrast, the low pay, long hours and strong attachment to local custom that prevailed across many rural areas of England proved to be powerful limitations on the adoption of seaside holidays by these communities. However, over the course of the second half of the nineteenth century, there is no doubt that as real wages increased and incidence of industrial holidays became more widespread, so the levels of access to the coast improved significantly for working-class people.

As a result, the later Victorian period is characterised by a clear trend towards the differentiation of resorts according to their social tone and character – the noisy, popular working-class resorts such as Blackpool (with their amusements, funfairs, entertainment and cheap food and drink) contrasting with the more sedate and refined qualities of middle-class Eastbourne or Bournemouth. Where established middle-class resorts were subjected to working-class invasions, wealthier visitors might either become displaced to more exclusive suburbs within the resort (for example, from Brighton to adjacent Hove), or to new places that were further removed from centres of industrial population and where the effects of distance still served as a barrier (for example, the west of England and, increasingly, abroad). Alternatively, wealthier patrons would confine their visits to times of the year when ordinary visitors would be absent.

The regulation of the social tone of the resort and wider questions of how seaside towns were developed and promoted during the latter part of the nineteenth century and the early twentieth century owed much to local patterns of landownership and the roles that municipal authorities opted to play. Leasehold systems enabled landowners to retain quite direct controls over how land was developed, and where such powers were reinforced by municipal regulation of behaviour (for example, via by-laws to control drinking) and investment in particular kinds of public facilities, then the social tone and character of a resort could be carefully controlled (Towner, 1996; Walton, 1983). Bournemouth provides an especially good example of how the refined character of the resort was consolidated by a combination of restrictive landownership and municipal investment in a particular range of leisure facilities that reflected the tastes and preferences of a middle-class clientele (Roberts, 1983; Soane, 1993). In contrast, the development of Blackpool, where landownership was fragmented amongst many freeholders and where the municipal authority took a contrasting view of its role in actively promoting a popular form of seaside leisure, resulting patterns were very different (Walton, 1983, 1994).

The further consolidation of resorts as leisure places during the twentieth century was promoted by several factors. Towner (1996) draws attention to the emergence of an almost universal demand for enhanced living conditions (including holidays) amongst the ordinary classes following the First World War. Improvements in living standards and significant increases in the incidence of paid holidays (eventually made a legal obligation of employers in Britain after 1938) reinforced popular demand, as did the emergence of lower-cost forms of holiday centred around camping and holiday camps. Active and competitive promotion of resorts and the complementary development by the railway companies and, later, bus companies of comprehensive and affordable holiday travel were additional factors in consolidating the mass markets for seaside leisure up to 1939.

Thus, by 1937, an estimated 15 million holidays of at least a week were taken in Britain. What proportion of these were at coastal resorts is uncertain but the capacity of the resorts to attract staying and day visitors was considerable. In the same year, for example, Blackpool drew some 7 million visitors and Southend over 5 million (Pimlott, 1947). In the post-1945 era, domestic tourism continued to develop to reach a peak of 40 million holidays in the early 1970s, most of which was directed towards seaside resorts or their hinterlands (BTA, 1995).

By 1945, the geographic distribution of resorts had also changed appreciably from the initial patterns of colonisation. In Britain, the original focus upon small, elite resorts, primarily on the coasts of Kent and Sussex, had evolved into a spatially extended pattern in which tracts of land along the coasts of southwest Scotland, North Wales, Lancashire, Yorkshire, East Anglia, Hampshire, Dorset, Devon and even remoter areas such as Cornwall, had been developed for tourism (Figure 2.3). Goodall (1992) has characterised this diffusion of resorts as a process of displacement from initial cores of development to successively more distant parts and as Towner (1996) shows, the patterns

SPAS, SEA BATHING AND THE GROWTH OF COASTAL RESORTS 35

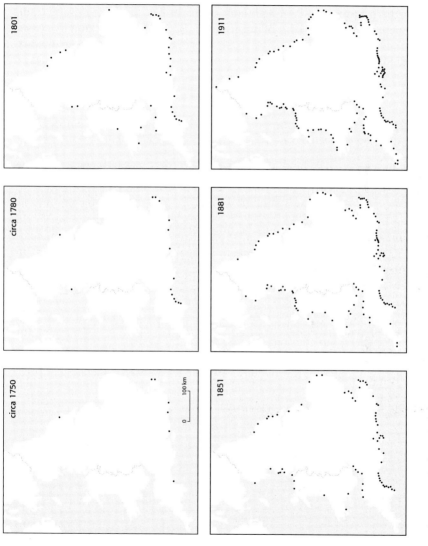

Figure 2.3 The spread of seaside resorts in England and Wales, 1750–1911
Source: After Towner (1996).

of spread that are evident in the British case were commonly replicated in developments of coastal resorts in Europe and North America. In France, for example, peripheral areas such as the French Riviera and Brittany had begun to develop from the 1830s, albeit for rather different reasons. On the Riviera, development came initially through processes of social displacement, as the area became a fashionable winter retreat for wealthy and aristocratic visitors from northern Europe and major French cities. In Brittany, in contrast, resort development was more a result of the gradual discovery of the region by artists, writers and travellers – attracted by its picturesque landscapes and the distinctive Celtic history and culture.

Resorts as tourist places

But to what extent were (and are) the seaside resorts purely formed as places of tourism? A primary objective of this section of the chapter is to question the notion that resorts may be simply interpreted as places shaped by tourism and the practices that are specific to tourism (see Chapter 1) and to argue instead that the nature of resorts can only be fully comprehended when the considerable influences of contemporary recreational practices are also taken into account. Walton (1983) suggests that resorts are distinctive primarily because of the *scale* at which leisure facilities are provided rather than because of significant differences in respect of *what* is provided or how it is used. The seaside resort does offer certain resources that are uniquely a part of seaside tourism – especially, of course, the beach and, to a lesser extent, the promenades and piers, but the contention offered here is that the manner of their use by the tourists and visitors is still primarily a product of routine recreational patterns. These patterns may, in their temporary transference to the seaside, become intensified and magnified, but they are otherwise clearly derived from more general models of recreational (rather than tourist) behaviour.

In developing this argument, the discussion will initially adopt a historic perspective in considering the influence of contemporary recreations on resort development in nineteenth-century England. A number of interrelated ideas will be presented, including: the incidence of local recreational practices that pre-date the formation of resorts; the transference of leisure cultures from places of normal residence to the seaside; and the influence on resort development of contemporary recreational tastes.

The incidence of local recreation

It is often erroneously assumed that the seaside was 'discovered' by the English aristocracy, but as several authors have emphasised, in many coastal locations popular local sea bathing cultures had existed quite independently of – and prior to – the practices of elite groups of outsiders (see, *inter alia*, Walton, 1983; Corbin, 1995; Towner, 1996). Moreover, local forms of sea bathing appear generally to have been conducted as popular recreations rather than for the reasons of health that first tempted the cultural elites to visit Brighton, Dieppe or Doberan. Corbin (1995), for example, notes how sea bathing formed a part of annual festivals in the Basque region of northern Spain, whilst along the Mediterranean shores of Italy, peasant communities routinely bathed for pleasure. Walton (1983) notes similar practices on the coast of Lancashire. Critically, however, such local customs did not create the need for any forms of organised development since at this stage the habit of sea bathing had yet to attract the staying visitor. Consequently, the investment in public facilities, hotels, boarding houses and summer homes that became an essential ingredient in the making of the new resorts in the latter part of the eighteenth century, naturally emerges as an innovation that draws attention to itself as an outwardly new form of behaviour but, in so doing, tends to obscure its precursor.

The failure to recognise the local recreational dimension has sometimes been a weakness in studies of resort development. In fact, resorts – both as places of fashion and, in the nineteenth century, places of rapid urban growth – quickly attracted permanent resident populations. Soane (1993) highlights the manner in which fashionable resorts such as Bournemouth attracted wealthy retirement migrants from an early stage, and the presence of both a tourism industry and a resident population encouraged the development of a permanent local service sector. Some resorts even had significant manufacturing roles (for example, Brighton possessed a large railway works). Consequently, local recreational practices and those of the visitor would have coexisted on a routine basis, and in many cases local investment in facilities often pursued provision that was compatible with the leisure activities of visitors and residents alike.

The transference of leisure cultures

When the aristocratic forms of seaside leisure developed from the late eighteenth century onwards, an important influence upon the organisation and character of those leisure patterns were the existing practices that were already established in the spas, the country houses and the royal courts. In eighteenth-century England, for example, modern distinctions between work and leisure as shared areas of experience were, of course, relatively meaningless. As members of an aristocratic elite, the patrons of the spas and resorts enjoyed a lifestyle that was essentially leisured and their routine recreations revolved around social gatherings and visiting; patronage of assembly rooms (for dancing or entertainment); the theatres, the reading and gaming rooms; and promenading. Significantly, though, the advent of a new set of leisure places at the seaside produced few (if any) discernible changes in these established recreational behaviours. As we have already seen, the seaside resort did not supplant the inland spa, it simply added a new set of fashionable venues to the annual social round. The imitative physical form of the early seaside resort confirms

Figure 2.4 The beach, the hotel and entertainment – three key elements in the Victorian–Edwardian middle-class seaside holiday (Llandudno, North Wales)

context of the discussions within this book it is, however, a moot point as to whether such excursion traffic constituted tourism or recreation. As day visitors, the excursionists fail to qualify as 'tourists' according to the criterion that defines tourists as staying away from home, yet by mixing freely with staying visitors in the resorts and by using many of the same facilities and attractions, they certainly engaged with the practices of the tourist and doubtless drew on the same set of experiences. It has also been shown that the influx of trippers, unwilling to modify their behaviour, induced instead the modification of the landscapes and structures of some resorts. This occurred especially on the routeways that linked the railway stations with the seafront, which would be rapidly colonised by the services and entertainments that the excursionists expected (Towner, 1996). More significantly, the development of the excursion trade accelerated processes of social segregation, not only within but also between resorts, helping to differentiate places according to their social tone (Urry, 1990a).

Alongside the growth of excursion traffic, resorts were also influenced by developments in popular entertainment. Cunningham (1980), in charting the changes in popular entertainment that affected urban–industrial Britain in general, notes several significant developments. These included:

- popularisation of several forms of theatre (for example, melodrama and pantomime);

- revitalisation of travelling shows, which had a lengthy tradition in popular leisure, but which in the mid nineteenth century became more sophisticated and developed new lines of entertainment – such as conjuring, Punch and Judy, menageries and early forms of circus;
- development of new forms of musical entertainment, including brass band competitions which date from the 1840s and, particularly, music halls which grew in popularity from the early 1850s;
- the revival of fairs from about 1850, especially as new technology and investment produced better rides and mechanical attractions.

These organised forms of entertainment immediately became integral to the tourist experience of the seaside, but in the context of the present argument, Walton's observation that 'the form and content of the shows and spectacles had been developed elsewhere – in London and the great industrial towns' is critical (Walton, 1983: 157). The resort was unusual in the concentration and scale of entertainment that it was able to provide, and it is true that in some areas, especially in the development of fairground amusements and rides, seaside resorts did become important innovators (particularly in the USA). In general, though, the seaside was 'more important as a carrier and concentrator of existing trends and fashions, than as an initiator of new ones' (Walton, 1983: 157). In other words, the character of tourist experience was shaped and directed by recreational practices developed elsewhere.

It is also instructive to observe how popular entertainment rapidly colonised some of the spaces and facilities that were unique to the seaside. This is especially true of seaside piers. These had originally been developed (from around 1825) as points of embarkation or landing of goods and passengers, although the novelty of being able to promenade above the sea attracted leisurely use from the outset. By the last quarter of the nineteenth century, however, the influence of popular recreational tastes had seen most seaside piers become centres of commercial entertainment, adorned with pavilions, theatres, concert halls, restaurants, penny arcades and sideshows (Walton, 1983).

Tourism, recreation and the seaside – post-1945 developments

By the start of the twentieth century, a pattern of resort-based tourism had become firmly established in Britain, Europe and the USA, whereby millions of people became regular visitors to the seaside towns – places in which a distinctive blend of tourist and recreational practices were followed, almost by ritual. As was noted above, the habit of seaside holidaymaking continued to develop throughout the first half of the twentieth century, to reach a peak of popularity in countries such as Britain by the end of the 1960s. Thereafter, however, the story has been rather different as many of the traditional resorts have begun to encounter varying forms of decline, and just as previous phases of growth were partly shaped by trends and tastes in both elite and popular forms of recreation, so we must again look for some of the explanations for subsequent decline in more recent shifts in recreational tastes and practices. In fact, it may reasonably be argued that as leisure patterns have developed during the post-1945 period, so the distinctions between conventional seaside tourism and ordinary recreation have become even more uncertain.

In pursuit of this argument, several trends are worthy of examination, including:

- the links between new patterns of leisure consumption, changing recreational tastes and the growth of alternative forms of tourism;
- the loss of a distinctive dimension to resort tourism and the continued convergence between leisure provision in resorts and other urban places;
- the changing relationship between resorts and their hinterlands;
- the return of resorts to a role as day attractions;
- the development of new coastal leisure sites.

New patterns of consumption and alternative forms of tourism

We have seen throughout this chapter that patterns of tourism and recreation are essentially reflections of wider social/cultural values, practices and structures. Culture and society are not, though, fixed entities – they evolve and as they do so, they produce new patterns in the production and consumption of recreational or tourist places. In recent years it has become clear that a particularly influential set of changes have been at work – variously designated as 'post-industrial', 'postmodern' or 'post-Fordist' (Williams, 1998). These have produced different patterns of recreation and tourism but, since these have tended to centre upon new activities and (usually) new places, they have begun to compromise the established role of seaside resorts as the pre-eminent centres for tourism and recreation.

Urry (1995) characterises the post-Fordist patterns of tourist consumption in terms of:

- a rejection of some mass forms of tourism and increased diversity of preference;
- fewer repeat visits and a proliferation of alternative attractions;
- increased turnover and change in the popularity of sites in response to fashions;
- associated growth of alternative forms of tourism.

These transformations, it is argued, have prompted enhanced levels of interest in, *inter alia*: cultural tourism; heritage and industrial tourism; adventure tourism; sport tourism; eco- or nature tourism and tourism centred on theme parks.

Such developments have inevitably refocused the tourist gaze onto new destinations in both the city and the countryside, but most of these new forms continue to perpetuate the close linkages between recreation and tourism that this chapter has argued are important. Theme parks, for example, can be seen as elaborations of the older fairs and amusement grounds that have formed a basis for popular recreation since the nineteenth century. Sport tourism (for example golfing, sailing or fishing holidays) is normally a direct extension of recreational participation and armchair viewing of sport on television. Many types of cultural and heritage tourism reflect and are informed by recreational reading, television viewing and hobby interests, whilst locations for popular television programmes have become major destinations for recreational and tourist excursions – often in unlikely places. For example, Davidson and Maitland's (1997) study of tourism to Bradford Metropolitan District in the industrial heart of northern England illustrates several of these connections. More than 1 million people annually visit Haworth village – the home of the Brontë sisters (reading); over 700,000 visit the National Museum of Film, Television and Photography (film and television); and a similar number visit Esholt – the setting for the TV soap opera *Emmerdale* (television). The replication of such patterns at a growing number of locations in Britain, Europe and North America – especially when reinforced by preferences for other new activities such as recreational shopping – seems to offer, therefore, a significant redefinition of tourism and recreation practices and their spatial locations, a redefinition that must have implications for traditional seaside resorts.

The loss of distinctiveness

In the nineteenth century, resorts developed as leisure places because they were distinctive attractions that offered a particular set of opportunities for the pursuit of pleasure. The special quality may have owed more to the unrivalled concentration of leisure facilities and their relative freedom of access than to the unique nature of what was provided (with the important exception of the sea itself), but the resorts were still distinctive places.

Over the last three decades of the twentieth century, however, the special character of the seaside resorts diminished significantly – a trend that was both a consequence and a cause of the reduced attraction of these traditional leisure places. The seeds of change lay in both the shifting tastes of popular leisure and the diversification of resort economies. The development of sophisticated and exciting forms of alternative leisure and recreation of the kinds outlined above undermined the rationale of seaside resorts. Shields (1990: 73) writes that 'for most Westerners in the late twentieth century, it is no longer necessary to create marginal zones, such as the seaside was, for reckless enjoyment'. Such opportunities have become routine parts of most recreational lifestyles. Urry (1990b) extends the argument by highlighting how the deindustrialisation of urban centres in the post-industrial era has eroded much of the sense of contrast between cities and resorts. Many urban places have developed extensive leisure provision of their own that is often superior to that found in older resorts and increasingly, as we shall see in Chapter 4, cities are themselves becoming major tourist places. This process of erosion began as long ago as the 1930s when a shift towards a less formalised, commercially oriented approach to seaside leisure saw the widespread construction in resorts of normal urban entertainments, such as cinemas and later bingo halls (Soane, 1993). Subsequent change has only served to reinforce the sense of convergence between towns and resorts in respect of their leisure provision.

The significance of local recreation provision in the changing basis of resorts has been amplified as resort economies have become more diverse and the resident populations (with their attendant leisure demands) have increased. One of the key trends that has prompted the growth of permanent populations in coastal resorts has been retirement migration. The links between retirement and the seaside are long established. Soane (1993) shows, for example, how some of the earliest residents of nineteenth-century Bournemouth were retired people, typically from a government or military background, and subsequent trends have reinforced the practice as improved standards of living have widened the social access to seaside retirement. Although subject to fluctuation according to prevailing economic conditions (Stillwell *et al.*, 1991), the concentration of the elderly through retirement migration to selected coastal areas was a significant feature of population change between 1920 and 1970 (Allon-Smith, 1982) and rates of movement increased still further in the 1980s (Stillwell *et al.*, 1993). As an example, Table 2.1 illustrates the overconcentration of the elderly within coastal districts in parts of southern England.

Table 2.1 The concentration of the elderly in coastal districts containing resorts – southern England, 1991

County/districts	% of population of pensionable age
Devon	23.6
East Devon	31.5
Teignbridge	26.0
Torbay	27.9
Dorset	26.0
Bournemouth	28.9
Christchurch	34.6
East Sussex	26.4
Brighton	21.8
Eastbourne	32.0
Hastings	23.3
Hove	27.3
Average for England and Wales	18.8

Source: Based on information from OPCS (1992).

The impacts of retirement migrants on coastal resorts have been quite diverse. Law and Warnes (1982) note that one of the more noticeable effects is upon the visible landscape of the resort and its vicinity, as predominantly owner-occupier groups of elderly migrants create active demand for housing, in particular bungalows. This may create a process of overspill development into adjacent coastal areas, as a result of which resorts may begin to lose their sense of spatial identity.

More importantly, perhaps, the presence of a large, elderly element within a resort population may hasten the demise of the resort itself. Cooper (1997) comments that retired elements are often openly hostile towards tourism. Retired people tend to perceive the industry as bringing few benefits for them personally, but considerable inconvenience through problems such as noise and congestion during the season. In a different vein, the presence of a large elderly population can adversely affect the image of a resort and discourage younger people from visiting places that are perceived as staid and relatively undynamic.

However, seaside resorts have not merely become places of retirement. Meethan's (1996) study of Brighton, for example, reveals how the tourist industry that formed the original stimulus to the growth of the town has become a diminishing element in a diverse urban economy that has seen significant recent development as a commuter settlement (for London), a regional centre for commercial services and higher education, as well as a place of retirement. In these circumstances, therefore, it is the recreational demands of the permanent population rather than those of the visitor that now shape the patterns of leisure provision in the resort. Soane (1993) offers another example of changing emphases – from Los Angeles – where the beach resorts have been supplanted as key tourist attactions by the new leisure industries centred upon Californian theme parks and the entertainment business of Hollywood. The beaches that formed the original attraction to tourists are widely accepted (and used) as part of the normal lives of the local residents, but for tourists the attractions now lie elsewhere.

The changing relationship between resorts and their hinterlands

This shifting focus of the tourist gaze has also been a contributory factor in the changing relationship between resorts and their hinterlands. In the period up to perhaps 1939, when the railway was the dominant means of holiday travel, local road excursions beyond the resorts were limited to a privileged few and most resort visitors were confined to the resorts themselves. Excursions by bus from resorts began to be popular in the 1950s, but it has been the revolution in personal holiday transport associated with car ownership that has exerted the greatest impact (see Figure 2.5). The new levels of personal mobility afforded by the car have generally altered the role of seaside resorts to a position in which they become bases for a much wider exploration of a hinterland that

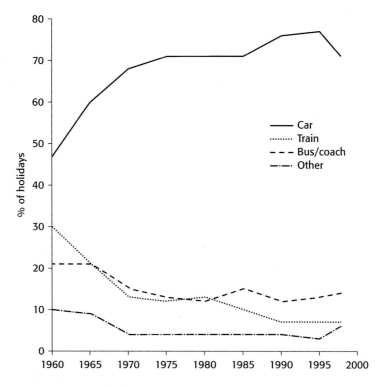

Figure 2.5 Changing levels in the main mode of holiday transport in the UK, 1960–98
Sources: Based on information from BTA (1975, 2001).

will include other resorts or urban centres, as well as accessible countryside. Marketing a modern resort, therefore, is partly concerned with promoting the attributes of the place itself and partly those of other places that are accessible from the resort – a shift in emphasis from local to regional resources. Even those who are not car owners may take advantage of these widening fields since coach operators in resorts remain as important providers of such services – routinely selling day excursions to a wide and diverse range of places and attractions. Figure 2.6 illustrates the visiting hinterland of the resorts of Torbay in south Devon, as defined by a sample of coach operators based in Torquay and Paignton (see also Figure 2.7). This shows that regular excursions not only visit other local seaside resorts and attractions such as Dartmoor National Park, but also encompass more distant tourist places in Cornwall and north Devon, as well as urban destinations as far afield as Bath, Bristol, Cardiff and Wells. Regular excursions also run to London and to Alton Towers theme park in Staffordshire (a round trip of more than 400 miles), although in this instance demand patterns might reflect recreational demands of local residents, as well as those of visitors to the resort.

TOURISM, RECREATION AND THE SEASIDE – POST-1945 DEVELOPMENTS 47

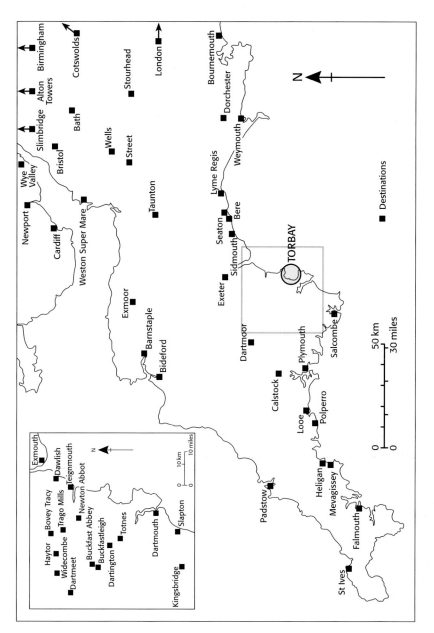

Figure 2.6 Day excursion destinations served by coach operators in Torbay, south-west England

Figure 2.7 The historic naval town of Dartmouth (south Devon) was never actively developed as a resort, but has become a popular attraction for visitors from neighbouring resort areas in Torbay

In developing these new patterns of resort-based tourism, a significant factor has been the growing appeal of rural forms of recreation and tourism. Once again, there are important recreational influences at work here for, as we shall see in Chapter 5, the development of rural tourism grew partly out of early patterns of recreational use of the countryside. From the late nineteenth century, when both walking and cycling first became popular and (in the case of cycling) affordable forms of recreation, day trips and excursions to local countryside sites and attractions formed essential ingredients in recreational lifestyles of lower-middle-class and working people in industrial towns and cities (Towner, 1996). Such excursions were often organised and actively promoted by newly formed recreational or sporting clubs, as well as older institutions such as Sunday schools or working men's associations and, in some instances, were sufficiently attractive to stimulate the further development of inland resorts. The resort of Windermere in the English Lake District, for example, had initially developed in the late nineteenth century as the centre of a genteel, upper-middle-class holiday trade based on enjoyment of scenery and outdoor activity. By the 1920s, however, the growth of excursion traffic (by train and charabanc) from the industrial cities of Lancashire and Yorkshire introduced new groups of recreationalists, keen to participate in the enjoyment of the scenery, countryside walks and water-based recreations (O'Neill, 1994).

In the post-1945 era this tradition has developed and been extended to the point at which recreational trips to the countryside are routine elements in many leisure lifestyles. According to recent visitor surveys in the UK, as many as 1.4 billion day trips are made to countryside sites annually (SCPR, 1997) and since such trips now constitute a familiar form of recreational behaviour, they also form a natural extension to normal tourist activity. As a result, where such sites fall within the hinterlands of resorts and other towns or cities, a natural coincidence between the trip patterns of tourists and those of local recreationalists will result, producing extended regions in which recreation and tourism become completely intermixed. Under such conditions it is arguably impossible to separate the two forms of behaviour, or the meanings and values that participants ascribe to their experiences.

For example, Dartmoor National Park in south-west England is directly accessible to significant concentrations of permanent population in the cities of Exeter and Plymouth (together with smaller urban centres such as Newton Abbot), but it is also a target for tourists staying in the major coastal resorts of Torbay and the other holiday areas of south Devon. As a result, visitor patterns in the national park are composed of three distinct elements: day visits by local residents (46 per cent); day visits by tourists staying elsewhere in the region (42 per cent); and a much smaller level (12 per cent) of visits by people staying within the park itself (Countryside Commission, 1996). It is probable that the spatial and temporal patterns of use of the two main groups show some differences, with a greater level of use by local people in periods away from the summer peak season, but the ubiquitous character of modern tourists suggests that coincidence and overlap of different leisure groups are inevitable.

The return of resorts to day attractions

The revolutions in personal transport that have promoted the role of resorts as bases for wider patterns of exploration, have also helped to erode their position as places at which people stay (and therefore, by some definitions, become tourists). Notwithstanding the growing problems of traffic congestion, the enhancement of road systems through the general development of motorways, dual carriageways and trunk routes has significantly increased the speed of travel and reduced the frictional effects of distance. Major tourist destinations therefore become accessible for short-duration recreational trips, especially day visits, in ways that would not have been possible prior to the end of the 1960s.

Changes in personal mobility, when linked with other trends (for example, towards increased levels of foreign holidaymaking or a pattern of tourism in which several shorter breaks have replaced the traditional, single holiday – see Table 2.2), have generally propelled conventional resorts into the role of short-term attractions, often at off-peak times of the year. In many British resorts, the peak season in July and August (see Figure 2.8) is increasingly dominated by a contracting and residual market of elderly and less affluent visitors that adhere to familiar patterns (often through a lack of alternatives), reinforced by a growing proportion of day trippers.

Table 2.2 Level and frequency of holidaymaking by UK residents (4 or more nights away), 1971–98

	Percentages						
	1971	1975	1980	1985	1990	1995	1998
Taking 1 holiday	44	44	43	37	36	35	31
Taking 2 holidays	12	14	14	14	15	17	16
Taking 3+ holidays	3	4	5	6	7	10	10
Total taking holidays	59	62	62	57	58	62	57
No holiday taken	41	38	38	43	42	38	43

Source: Based on information from BTA (1995, 2001).

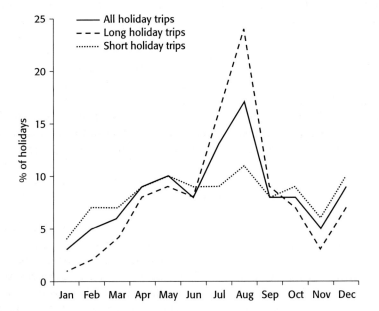

Figure 2.8 Seasonal pattern of holiday-taking in the UK, 1999
Source: Based on information from BTA (2001).

Data to illustrate the contracting temporal patterns of resort use are difficult to locate but although it may be fragmentary, the evidence is still persuasive:

- According to the BTA (1995), nearly one-third of holiday trips in the UK are now of two nights' duration or less.
- The UK Day Visitors Survey (SCPR, 1997) claims that an estimated 198 million day trips were made to the British seaside in 1996 – a figure that, if accurate, would mean that staying visitors are outnumbered by recreational trippers by as much as 15 to 1.

- During the 1980s, British seaside resorts lost an estimated 39 million visitor nights (21 per cent of their market) and as a consequence, many resorts saw significant losses of (particularly) hotel and guest house accommodation through closure (Cooper, 1997).

Such trends, when taken together, point firmly towards the conclusion that in countries such as Britain, at least, the contemporary resort is as much a recreational place as it is a centre of traditional forms of tourism.

The development of new coastal leisure sites

Finally, it is worth noting the increasing role of new coastal leisure sites. Coastal development outside resorts is not, of course, a new phenomenon, as the growth of holiday camps illustrates. As early as 1939, government and bodies such as the Council for the Protection for Rural England were already expressing concerns over encroachment by tourism onto undeveloped stretches of coastline and urging that development of holiday camps, in particular, should be precluded from certain categories of land. These included high-quality farmland, unspoilt coastline and mountain or moorland areas (Ward and Hardy, 1986). Nevertheless, the relative weakness of the statutory planning systems in the UK prior to 1947 meant that by the end of the Second World War, extensive stretches of coastline beyond the established resorts had already been affected by unplanned and haphazard development of holiday camps (Figure 2.9). On the Lincolnshire coast alone, there were 51 camps with a visitor capacity of 32,000 and which ranged in size up to a maximum of 37 ha at the Butlin's camp at Skegness – to all intents and purposes, self-contained resorts in their own right (Ward and Hardy, 1986).

A similar tale is evident in the development of caravan sites. Many of the early holiday camps also made limited provision for caravans, but the growing popularity throughout the post-1950 period of this relatively cheap and flexible form of accommodation (see Table 2.3) has increased the demand for sites. Pryce's (1967) study of caravans in Wales (an area that has been particularly affected) shows how site developments not only extended the urban areas of established resorts such as Rhyl and Prestatyn, but were also in the forefront of colonisation of previously undeveloped coastal zones in areas such as Pembrokeshire and along the shores of Cardigan Bay.

More recently, however, other forms of development have contributed to the process. In Britain, France and Spain, for example, conventional resorts now form but one element in a pattern of coastal leisure development that has seen new combinations of tourism, recreation and local leisure. One of the most conspicuous forms of recent coastal development has been the provision of marinas. The demand for marinas has largely emanated from the growth in participation in recreational sailing. According to Smith and Jenner (1995), greater affluence and the relative reduction in the cost of boat ownership have been responsible for significant increases in demand that, by the early 1980s, were fuelling a period of active marina development. Sidaway (1991) suggests that

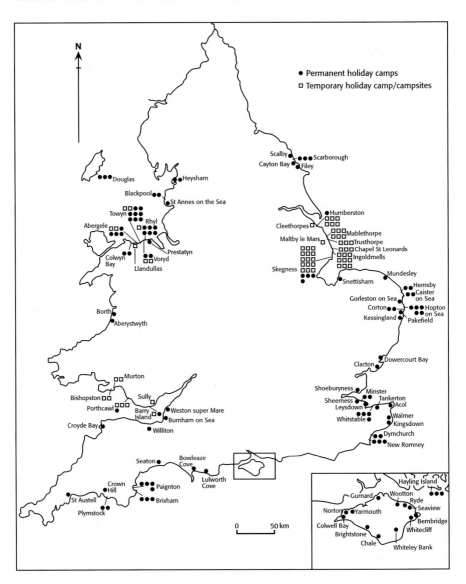

Figure 2.9 Distribution of holiday camps in England and Wales, 1939
Source: After Ward and Hardy (1986).

in 1989, Britain had 370 coastal marinas and moorings with just over 75,000 berths, most of which were concentrated along the southern and south-eastern coasts of England. By 1995, Britain lay second only to Spain within Europe, in the number of marinas that it had developed (Smith and Jenner, 1995).

These marinas have taken a variety of forms. Some are relatively simple moorings in established recreational zones such as the River Hamble on the

Table 2.3 Accommodation used on main holidays in Great Britain, 1955–98

	Percentages					
	1955	1965	1975	1985	1993	1998
Licensed hotel	14	13	15	19	21	25
Unlicensed hotel/boarding house	27	28	12	6	6	4
Friends and relations	31	25	26	23	17	18
Caravans	8	13	21	21	20	24
Rented accommodation	7	8	11	13	15	14
Holiday camp/village	4	6	6	9	10	9

Sources: British Travel Association (1969); BTA (1995, 2001).

Solent (Goodhead *et al.*, 1996), but as the number of river moorings has reached its limit, a more typical pattern for new marina construction has been the mixed waterfront development that combines leisure facilities with housing and, in some of the more ambitious schemes, retailing and commercial space as well. These new forms (that took their inspiration from highly successful integrated developments on the French Mediterranean coast such as Port Grimaud) have sometimes taken coastal recreation and tourism into the heart of major industrial cities. St Katherine's Dock in London was one of the first examples of comprehensive regeneration of former dockland in the UK (with a marina, hotels, apartments, retailing and commercial space – Figure 2.10), although the scope for this kind of regeneration had already been demonstrated elsewhere, for example at Baltimore in the USA. Other sites, such as Liverpool's Albert Dock, soon followed. Figure 2.11 illustrates another example of marina-based waterfront development at Swansea's South Dock. This is a smaller-scale version of the type of redevelopment that has taken place at Liverpool in which disused docks have provided a focus for regeneration of a decaying environment. Apart from the moorings for recreational sailors, onshore development of a leisure centre, museum, shops, bars and restaurants have reinforced the importance of South Dock not just as a tourist site but also as a local leisure space, as well as a residential zone. Surveys suggest that half of all visitors to this 'maritime quarter' come for informal leisure purposes (Edwards, 1996).

Port Solent (Figure 2.12) provides a rather different example. This scheme is a totally new development, privately funded and planned as an integrated complex of a marina, houses, apartments, office premises, retailing, restaurants, bars and a 10-screen multiplex cinema. The location is a former rubbish dump in the north-east corner of Portsmouth Harbour – a site that prior to development would have scarcely been envisaged as a leisure setting – yet by 1991, it was drawing over 200,000 visitors per annum, both recreational users from local communities and tourists from further afield (Edwards, 1996).

Figure 2.10 Part of the redevelopment of St Katherine's Dock, London. The former warehouses now contain offices, tourist shops, bars and restaurants, whilst the water space is dominated by pleasure craft

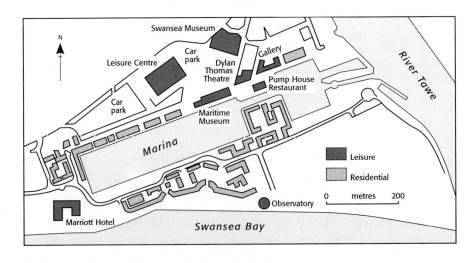

Figure 2.11 Recreational and tourist development at South Dock, Swansea

TOURISM, RECREATION AND THE SEASIDE – POST-1945 DEVELOPMENTS

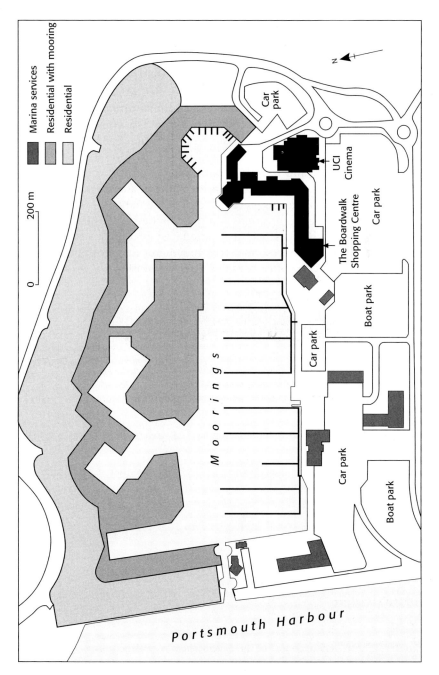

Figure 2.12 Port Solent Marina development, Portsmouth

Conclusion

This chapter has brought together a wide-ranging discussion in pursuit of what is an essential argument within the book as a whole. It has attempted to show that from the earliest stages of resort formation, the customs and practices that defined the resorts – first as elite spaces and then as centres of popular attraction – drew heavily upon established and routine forms of recreational behaviour in shaping the new patterns that we would come to define as 'tourism'. This symbiosis between tourism and recreation continued to influence the development of resorts as centres of mass recreation in the first half of the twentieth century and has also contributed to more recent processes of change and decline, as recreational tastes and preferences have altered. Throughout this period, it has been argued, the special nature of resorts owed as much to the unusual concentrations of opportunities for leisure that these places afforded, as to the presence of a unique set of tourist practices. To understand fully the nature of resorts therefore requires that we appreciate the nature and full extent of this often complex interrelationship.

Questions

1. What were the principal factors that shaped patterns of resort development in the UK before 1830?
2. How did innovations in transport affect resort development over the period between 1840 and 1939?
3. In what ways did resort development up to 1939 reflect popular recreational tastes and preferences?
4. Explain how recent shifts in recreational patterns have undermined the position of resorts as centres of tourism.
5. Examine the ways in which the development of coastal sites of recreation and tourism have become more diverse since 1950.

Further reading

There are several excellent accounts of the development of seaside resorts as centres of coastal recreation and tourism, including:

Soane, J.V.N. (1993) *Fashionable Resort Regions: Their evolution and transformation*, Wallingford: CAB International.

Towner, J. (1996) *An Historical Geography of Recreation and Tourism in the Western World, 1540–1940*, Chichester: John Wiley.
Walton, J.K. (1983) *The English Seaside Resort: A social history*, Leicester: Leicester University Press.
Walvin, J. (1978) *Beside the Seaside: A social history of the popular seaside*, London: Allen Lane.

Although first published more than 50 years ago, Pimlott, J.A.R. (1947) *The Englishman's Holiday: A social history*, London: Faber, remains a classic text and is well worth reading for insights into early phases of development.

Tourism, recreation and international travel

CHAPTER 3

Introduction

The coincidence of recreation and tourism spaces and practices within a domestic system is perhaps unsurprising, but in the realms of international travel there may be greater reason to anticipate a sharper sense of distinction between the two spheres of leisure. International travel generally imposes greater levels of spatial displacement of participants between home and destination; it invites a clearer sense of contrast in society, culture and environment; and it normally requires increased levels of expenditure that promote the personal significance of foreign travel over the more routine experiences of local recreation and tourism. However, as the taste for foreign travel has become widespread within westernised nations and increasingly prevalent in parts of the globe where such practices were previously less common – for example, in eastern Europe or South-east Asia – so there are signs that a synergy between recreation and globalised tourism is also emerging.

The chapter falls into two primary sections. First, the growth of international tourism over the second half of the twentieth century is outlined and some key spatial and structural changes in the pattern of global tourism are identified. Second, the factors that have promoted these developments are explained and illustrated in greater detail. The chapter concludes by returning briefly to the subject of convergence between recreation and tourism that was introduced in Chapter 1, highlighting those aspects in the preceding discussion that illustrate this theme.

The growth of international tourism since 1945

The World Tourism Organization (WTO) estimates that international tourist arrivals in 1999 numbered almost 665 million (BTA, 2001). This represents a 4 per cent increase on the previous year's total and a continuation of a seemingly irresistible trend that has seen average annual increases over the 10 years leading up to 1999 of the order of 4.5 per cent. Furthermore, the longer-term trends – as illustrated in Figure 3.1 – show that rates of expansion have generally tended to accelerate through time, with only temporary restraints on growth associated with global crises (such as economic recession in the mid-1980s or the Gulf War of 1991). Thus, between 1967 and 1976, the world market expanded by some 99 million arrivals; between 1977 and 1986 by 110 million; and between 1987 and 1996 by 256 million. Given these levels of change, by the early years of the twenty-first century international tourist arrivals worldwide will be exceeding 700 million per annum. This not only points to the importance of tourism within the leisure lifestyles of significant proportions of the population, but also to the growing centrality of such activities within economy and society, as more of the world's population experience the post-industrial shift from systems based upon production to those based on consumption (Urry, 1992).

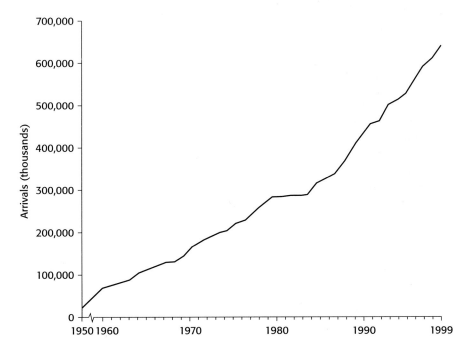

Figure 3.1 Growth of international tourist arrivals, 1950–99
Sources: Based on information from WTO (1995, 1998) and BTA (2001).

Table 3.1 International tourist arrivals by WTO region, 1986–96

	Millions								% share	
	1986	1990	1991	1992	1993	1994	1995	1996	1986	1996
Africa	9.3	15.1	16.2	17.8	18.3	18.7	19.2	20.6	2.7	3.5
Americas	71.4	93.4	96.5	103.4	103.6	106.4	110.4	114.7	21.1	19.3
E. Asia and Pacific	33.8	53.2	55.0	62.7	69.6	75.2	79.7	87.0	10.0	14.7
Europe	215.3	284.5	284.3	305.1	311.9	328.2	336.4	351.6	63.5	59.2
Middle East	6.4	9.0	8.4	10.5	10.9	12.1	13.7	15.3	1.9	2.6
South Asia	2.7	3.2	3.3	3.6	3.6	3.9	4.3	4.5	0.8	0.8
World	339.0	458.3	463.6	503.1	517.9	544.5	563.6	593.6	100.0	100.0

Source: Based on information from WTO, cited in Bar-On (1997).

The composite picture of the growth of international tourism conceals, however, important variations between regions and also masks the processes of spatial redistribution of visitors that are changing the balance of tourist movement. Table 3.1 sets out recent patterns of change, both in aggregate figures and in the market share held by different WTO regions. Several points are evident, but the dominance of Europe and, as a more recent development, the growth of long-haul forms of travel, invite particular attention.

Europe as an international destination

The dominance of Europe as a destination is a firmly established feature within the geography of international tourism. Three out of five tourist arrivals in 1996 were recorded in Europe, whilst most of the remainder were in the WTO American and East Asia/Pacific regions. Of the top 20 destinations in 1996, 14 were European, 3 were American (USA, Canada and Mexico) and 3 (China, Hong Kong and Thailand) lay in the East Asia and Pacific region (Bar-On, 1997). Amongst the top 10 generating countries, 7 were European, 2 were American (USA and Canada) and one (Japan) was East Asian. This pattern tends to confirm the view of Urry (1992) that global tourism is, for the present, largely confined to citizens of the West and certain parts of the Pacific rim, although as we shall see shortly, there is also evidence that participation in international tourism is diffusing into areas where such forms of travel were previously exceptional.

Although the European share of global tourism has fallen in recent years (see Table 3.1), absolute totals have continued to rise, moving from 282 million arrivals in 1990 to 360 million in 1997 (WTO, 1998). The significance of Europe in international travel patterns is a product of a complex series of inter-relationships. The WTO (1997) draw attention to a range of general influences that have helped to promote tourism. These include, *inter alia*:

- the maturity of its tourism industries and the extent of its tourism infrastructure;
- the established traditions in leisure travel amongst relatively affluent and mobile populations;
- a rich diversity of natural and artificial attractions;
- a geopolitical structure that, through the juxtaposition of a large number of relatively small states, naturally inflates the incidence of international travel within the region (Jansen-Verbeke, 1995).

Additionally, European tourism has benefited from more recent initiatives, including:

- the expansion of high-speed rail links (including new routes via the Channel Tunnel);
- the construction of major theme parks (particularly Disneyland Paris);
- the growing significance of business tourism within an expanding European Union.

As a consequence of the factors set out above, European travel patterns are dominated by Europeans. In 1996, for example, 80 per cent of international arrivals in Europe were from other European states – a level that has been consistently maintained and which reflects both a maturity and an associated stability within the market (WTO, 1998).

However, the spatial patterns of international tourism within Europe have shown some potentially significant shifts in recent years. Traditional patterns tended to be dominated by north to south movements, from a largely industrial–urban base in the cooler latitudes of northern Europe, to the warmer shores of the Mediterranean (Burton, 1994). These flows were supplemented by less substantial movements into secondary areas such as the Alps and a pattern of inter-city trips that owed part of their dynamics to the legacy of the Grand Tour, with their emphasis upon a comparatively small number of historic centres of culture, such as Paris, London, Rome and Vienna. More recently, though, patterns of tourist movement in Europe have tended to become more complex, in at least three distinct ways.

Development of the eastern Mediterranean

First, there has occurred a spatial extension in the locations of some of the popular forms of Mediterranean beach tourism from the west to the east of the region, particularly to newer destinations in Greece, Turkey, Cyprus, parts of the former Yugoslavia and Israel. This is partly a product of the positive attraction of new tourism places and the facilitation by the travel industry of successively longer journeys, which encourages spatial displacement of tourists within what Turner and Ash (1975) have styled 'the pleasure peripheries'. Less positively, though, changes have also been interpreted as a reaction to the emergence of a number of tourism-related problems in destinations such as the popular resorts of the Spanish Mediterranean. These have included environmental damage due to a general incidence of unregulated development, together with more localised problems of congestion, pollution, over-commercialisation, increases in rates of tourism-related crime and drunkenness, and loss of image (see, as an example, Pollard and Rodriguez, 1993). For the Spanish market, in particular, overdependence upon an undifferentiated, mass market has also been recognised as a potential weakness (Vera and Rippen, 1996) and whilst overall levels of tourism to Spain are still high (40 million visitors in 1996), comparison of recent growth rates suggests some loss of competitive advantage amongst destinations in the western Mediterranean. Between 1992 and 1996, for example, Spanish tourism grew by 11 per cent, but visits to Italy, Turkey and Israel grew by 26, 22 and 40 per cent respectively (WTO, 1998).

New patterns of tourism in eastern Europe

Second, the general collapse of Communism in the former Soviet bloc and the consequent relaxation of controls on travel by its citizens, has opened new opportunities for tourism within east European states and, to a lesser extent, for tourism between east and west Europe. The impact of these changes upon the geographic patterns of tourism are, however, highly variable. Table 3.2 shows changes in the volume of international arrivals in six east European countries. The data suggest that not only are there clear differences in the scale of tourism to different parts of the region (cf. Hungary, Poland and the Czech Republic with Bulgaria, Romania and the Slovak Republic), but also that some destinations – having witnessed periods of significant growth in the early 1990s – have now stabilised their visitor levels or even suffered apparent decline.

Under Communist regimes, tourism within this area had a strong domestic dimension and international visits were primarily generated from other east European states and contiguous areas. In some localities, this tradition has continued and strengthened. Tourism to the Black Sea coast of Romania, for example, has benefited from increased patronage by visitors from the Russian

Table 3.2 Changing levels of international arrivals in selected eastern European states, 1992–6

	Visitors (in thousands)				
	1992	1993	1994	1995	1996
Bulgaria	1,322	3,182	3,896	3,466	2,795
Czech Republic	10,900	11,500	17,000	16,500	17,000
Hungary	20,188	22,804	21,425	20,690	20,674
Poland	16,260	16,930	18,825	19,215	19,410
Romania	3,798	2,911	2,796	2,608	2,834
Slovak Republic	566	653	902	903	951

Source: Adapted from WTO (1998).

Federation and Ukraine (Light and Andone, 1996), whilst countries such as the Czech Republic and Poland have gained visitors through their proximity to west European markets, especially Germany. (Prague, for example, lies only 50 km from the German border and has become a popular and affordable destination for short-break tourism by Western visitors (Johnson, 1995).)

More widely, however, the collapse of Communism and associated transitions from a command to a market economy appear to have initiated a sharp decline in tourist visits between former Soviet bloc countries. In some instances, especially in Poland, the Czech Republic and eastern Germany, the more affluent citizens have begun to take advantage of new opportunities to travel to west European destinations previously denied them (Light and Andone, 1996). At the same time, others on whom much of the domestic and international tourism in eastern Europe previously depended, have been forced to curtail their leisure visits in the face of significant increases in local prices, reductions in the real value of wages, and much higher levels of unemployment that have followed in the wake of the introduction of free-market economies (see Bachvarov, 1997; Balaz, 1995; Johnson, 1995; Light and Andone, 1996). Consequently, much of the 'tourism' between states in eastern Europe that is recorded in Table 3.2 is currently undertaken as day visits for purposes such as the buying and selling of goods or even to seek employment, rather than as leisure travel.

The contribution to the general growth of European tourism that emanates from the east of the region is therefore uncertain and, in all probability, overstated by the data presented above. The tourist potential of the region (with its significant resources in historic cities, sites of culture and its diversity of landscapes) is considerable, but at present, flows of the more affluent west European visitors to the region are relatively small, and a period of economic adjustment and stability is evidently required in order to restore tourism markets within the east European states and permit future growth.

The growth of short-break international tourism

A third key change in European tourism is that new opportunities have arisen for shorter-range international travel, often in short breaks of three nights or less. In particular, the opening of the Channel Tunnel has contributed to an increase in the interchange of tourists between Britain, northern France and the Low Countries, as the following case study illustrates.

Case study: The impact of the Channel Tunnel on tourism from Britain to France

The opening of the Channel Tunnel in October 1994 induced a number of direct and indirect improvements to transport between Britain and France. The introduction of a new, high-speed rail link providing services for passengers (by

Eurostar) and cars with their passengers (by the Eurotunnel Shuttle) has established a new level of competition on the short crossing from Dover to Calais. This has challenged the traditional monopoly of the conventional ferry services and created new opportunites for short-stay tourism to the opposite coast. Not only has the rail link reduced surface journey times between Britain and France by a half, the Eurostar service between London and Paris provides direct competition with air services with a centre-to-centre timing that is 20 minutes faster than journeys using scheduled air services (Vickerman, 1995). Furthermore, the new link has cut the costs of journeys, stimulated restructuring (especially through the merger of the largest ferry operators P&O and Stena, and the closure of some smaller operators such as Sally Line) and forced improvements to the services of competing ferry companies. This has encouraged the introduction of high-speed ferries on several of the short crossings, increased frequency of sailings, as well as widespread price reductions, discounting and active promotion.

The consequence of these developments has been a significant increase in visits by UK citizens to Europe using either the Channel Tunnel or the competing ferry services. According to Mintel (1999a), tourist numbers rose from 10.4 million in 1994 to an estimated 15.3 million in 1999 – an overall increase of 47 per cent and accounting for a third of all passengers crossing the Channel. Within this expanding market, the share held by services through the Channel Tunnel has risen from just 1 per cent in 1994 to over 40 per cent in 1998. Projected figures to 2003 suggest that by then, the Tunnel will be carrying nearly 55 per cent of passengers (in all categories) making surface crossings between Britain and Europe (Figure 3.2).

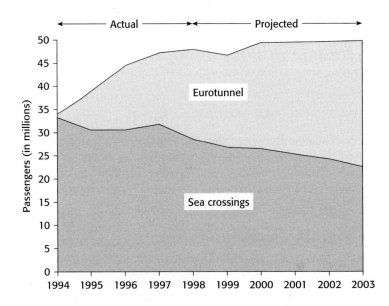

Figure 3.2 Actual and projected growth in cross-Channel traffic, 1994–2003
Source: Based on information from Mintel (1999a).

Whilst the bulk of this growing market is still accounted for by business travel – especially via Eurostar from London to Brussels and Paris – and the movement of freight by Le Shuttle and the ferries, a key component in this growth is due to developments within international tourism markets. Page (1999) argues that the promoters of the Tunnel failed to anticipate the ability of the route to stimulate new demand for tourist travel. This is surprising given that over one-third of cross-Channel visits are for holidays (Mintel, 1999a), but it soon became evident that as services through the Tunnel developed, so the traditional long-stay visits were being supplemented by an expanding base of day and short-break trips, as city-break tourism and as sightseeing and shopping excursions to the French coastal ports.

The growth of the day-visit/short-break market had been anticipated by tourism analysts (see Page and Sinclair, 1992), even if the operators themselves were slower to appreciate the potential. The realisation of that potential has been partly a consequence of the reduced travelling times that makes day trips between Britain and France feasible, but is also a product of the extended catchments for day excursions that the new, fast services have created. Residents of the Channel ports and their immediate environs have long been able to travel for short visits to the opposite coastline. However, the speed of the Eurostar/Euroshuttle services extends that opportunity across much larger areas of southern England and, in particular, to London.

Day trips to France had been given a significant boost with the introduction of the EU Single Market in 1993. This permitted travellers within the EU to import unlimited supplies of goods for their personal consumption, and the high levels of duty in the UK – especially on alcohol and tobacco – was fostering a growing market for shopping excursions to French ports prior to the opening of the Tunnel. Since 1994, data suggest that this market has more than doubled, from an estimated 2.7 million day trips in 1994 to 5.5 million in 1999, including a growing percentage who travel regularly and often (Mintel, 1999a).

The figures for cross-Channel travel also include some 3.1 million short breaks taken abroad by the British (Mintel, 1997b). Of these, 80 per cent were directed to France, Belgium and the Netherlands. Within short-break forms of tourism, two distinctive categories of destination have emerged strongly: first, city destinations such as Paris, Amsterdam, Bruges and Brussels; and second, visits to new forms of attraction, especially the major European theme parks.

International visitor levels to major cities are often substantial. Table 3.3 lists the major European cities according to estimates of the number of hotel nights spent by foreign visitors in 1992. This reveals not just the scale of this form of tourism, but also how it is currently concentrated into the major European capitals and centres of culture. It is interesting to note, however, that a little further down the list come cities such as Glasgow, Manchester and Birmingham, illustrating that the shifting focus of the tourist gaze is also beginning to draw cities with no particular history of urban tourism into the world of globalised travel.

Table 3.3 Foreign hotel nights in European cities, 1992

Rank	City	Nights (millions)	Rank	City	Nights (millions)
1	London	22.9	11=	Dublin	2.7
2	Paris	21.9		Florence	2.7
3	Vienna	7.5		Munich	2.7
4	Amsterdam	3.8	14	Barcelona	2.6
5	Budapest	3.7	15	Milan	2.4
6	Madrid	3.4	16=	Copenhagen	2.1
7	Prague	3.2		Lisbon	2.1
8=	Brussels	3.1	18	Istanbul	2.0
	Edinburgh	3.1	19=	Frankfurt	1.7
	Venice	3.1		Geneva	1.7

Source: Law (1996).

One of the key attractions in city tourism is the opportunity to indulge new tastes for leisure shopping. Tourists in many situations spend significant amounts of time on shopping, as part of a wider visit, but recent trends point to the emergence of new sectors of day and short-break international tourism where shopping is the primary purpose. This is evident in the cross-Channel trips of British tourists to buy foodstuffs and cheaper wines, beers and tobacco in France, but is also seen in other destination areas outside Europe. For example, a study by Timothy and Butler (1995) of cross-border shopping between the USA and Canada identified nearly 80 million day trips annually (of which three-quarters were from Canada to the USA) and nearly 19 million trips of up to two nights' duration. Growth has been encouraged by the factors that have promoted the expansion of foreign tourism in general (see below), but additionally, the preponderance of Canadians visiting the USA was attributed to local factors as well. These included competitive pricing of US goods; favourable exchange rates between the US and Canadian dollars; active promotion; and the availability of Sunday shopping in the USA.

Alongside the appeal of urban tourism and leisure shopping, short-break tourism has also been stimulated by the development of new commercial attractions – developments that have proven capable of rivalling and even exceeding established cultural and historic attractions in their levels of visitor attendance (see Table 3.4). Of course, in establishing and sustaining such levels, these attractions draw upon a significant number of day visits by local or regional populations, but the presence of national and international tourists at venues such as Disneyland Paris (which includes extensive provision of on-site tourist accommodation) is not to be neglected. To this extent, the development of new attractions provides an additional component in the geographical spread of leisure and tourism within Europe. As Jenner and Smith (1996) observe, with one or two notable exceptions, many of the established cultural and historic sightseeing attractions of Europe tend to be in the south, especially in Italy, whereas in contrast, most of the newer commercial attractions

Table 3.4 Attendance at selected attractions in Europe, 1995

Attraction	Visitors (thousands)
Disneyland, Paris	10,700
Notre-Dame Cathedral, Paris	10,000
Blackpool Pleasure Beach, England	7,300
Mont-St-Michel, France	7,000
Louvre, Paris	6,300
British Museum, London	5,746
Eiffel Tower, Paris	5,500
Tivoli, Copenhagen	3,800
Cologne Cathedral, Germany	3,500
Alton Towers Theme Park, England	2,707
Port Aventura Theme Park, Spain	2,700
Tower of London, England	2,537
Legoland Theme Park, Denmark	1,300

Source: Adapted from Jenner and Smith (1996).

are firmly located in the urban–industrial heartland of northern Europe, in Germany, France, Belgium and the Netherlands.

The growth of long-haul tourism

Whilst Europe and, to a lesser degree, North America continue much of their traditional dominance as destinations, their position is slowly being eroded as the spatial range of tourists increases and their patterns of preference alter. Although, as Figure 3.3 shows, in absolute terms the expansion of European markets continues to be consistently strong, the percentage share of the world market that is held by Europe shows a relative decrease. In contrast, emerging destinations in Africa and, particularly, in East Asia and the Pacific, have increased their share significantly. The East Asia and Pacific market grew by 157 per cent in the 10 years up to 1996 as annual arrivals climbed to 87 million (Bar-On, 1997). So, within the global growth of international tourism, there are also significant redistributions of activity taking place.

The development of these newer destinations is primarily the product of two factors: the increased appeal of long-haul forms of tourism and, in the specific case of East Asia and the Pacific, much wider incidence of intra-regional travel of the kind that is significant within Europe. Long-haul travel touches all parts of the global tourism market and Figure 3.4 illustrates the pattern of outbound long-haul tourism from Europe in 1996. This reveals the importance of the transatlantic links to North America and the Caribbean, but also shows significant flows of tourists to Africa, the Middle East and East Asia and the Pacific.

The impact of long-haul tourism upon regional markets is, however, highly variable (Mintel, 1997a). In Africa, the Middle East and, especially, South Asia, long-haul tourism contributes significantly to overall levels of visitor arrivals (Table 3.5) but elsewhere, levels are very much lower. This is interesting since

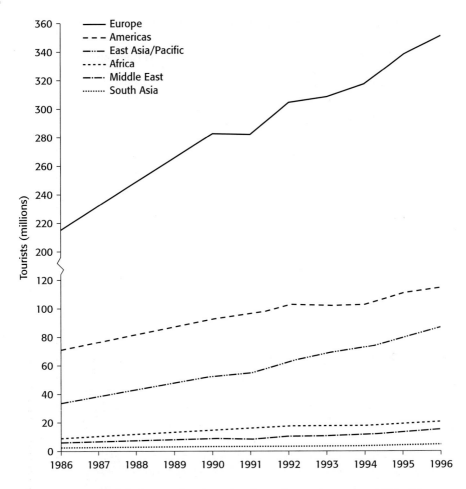

Figure 3.3 Growth in international tourism in major market areas, 1986–96
Source: Based on information from WTO (1998).

it shows that the rapid growth of tourism in the East Asia and Pacific region is far more dependent upon intra-regional movement of tourists than visitors from outside. The long-haul market to East Asia and the Pacific (principally from Europe and the Americas) increased from 8.1 to 18.1 million between 1985 and 1996, but the intra-regional market grew from 22.8 to 69.0 million over the same period (Table 3.6). The WTO (1997) offers a range of explanations for this growth, including:

- increased income and leisure time stemming from dynamic trade and investment conditions in several South-east Asian economies;
- active government promotion of tourism;
- greater levels of political stability;
- reductions on political restrictions on travel in some areas, especially China.

THE GROWTH OF INTERNATIONAL TOURISM SINCE 1945   69

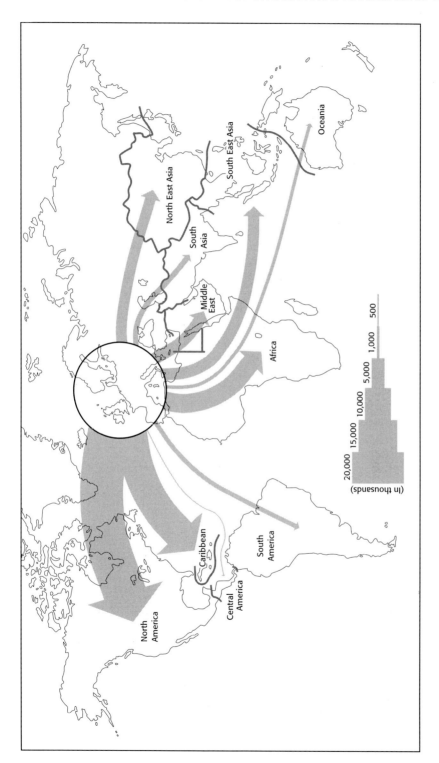

Figure 3.4 Pattern of outbound international tourism from Europe to WTO regions, 1996
Source: Based on information from WTO (1998).

Table 3.5 Market shares of long-haul and intra-regional tourism in WTO regions, 1996

	Percentages	
	Long-haul	Intra-regional
Africa	59.0	41.0
Americas	25.8	74.2
East Asia and Pacific	20.7	79.3
Europe	21.7	78.3
Middle East	56.7	43.3
South Asia	78.0	22.0

Source: Based on information from WTO (1997).

Table 3.6 Changing structure of the East Asia and Pacific region market, 1985–96

Origin	Tourist arrivals (thousands)		
	1985	1996	% increase
Intra-regional	22,781	68,978	+202
Europe	3,245	9,894	+205
Americas	3,623	6,009	+66
South Asia	840	1,356	+61
Africa	112	398	+255
Middle East	238	390	+64
Total long-haul	8,058	18,047	+124

Source: WTO (1997).

The growth of tourism to China has been particularly significant, with numbers increasing from 3.5 million in 1980, to 10.4 million in 1990 and an estimated 22.8 million in 1996 (WTO, 1997).

From the preceding discussion it is evident that any study of the geographical pattern of global tourism needs to take account of three primary processes:

1 The overall growth in the scale of the activity.
2 The spatial redistribution of tourism and the associated patterns of relative increase or decrease in levels of foreign tourism to individual regions or states.
3 The increased democratisation of international tourism, as is evidenced in, for example, widening participation in tourism by residents in eastern Europe and, especially, South-east Asia.

In outlining these processes, passing reference has already been made to some of the reasons that underpin the changes that are taking place, so to develop these themes, the discussion now moves to consider the factors that are promoting the globalisation of tourism in greater depth and breadth.

Factors promoting global tourism

The factors that have encouraged and enabled the development of international forms of tourism are many and varied. For convenience, therefore, the discussion is arranged under four sub-headings – the development of the travel industry; the impact of technology; the impact of economic and political convergence; and lifestyle changes – each of which attempts to summarise groups of key themes.

The development of the travel industry

Tourism, in all its guises, is dependent upon a supporting infrastructure of facilities and services but it is clear that the large-scale development of international tourism, in particular, would not have occurred without the simultaneous development of an accessible and supportive travel industry. This has provided not just the basic elements of accommodation, transport and local entertainment but, more influentially, has developed new structural forms of travel – especially in packaged tourism. Through the application of professionalism, the industry has also brought a level of flexibility, sophistication and simplification to the provision and promotion of international tourism that has largely eliminated many of the risks and difficulties – both real and perceived – that were once attached to foreign travel. Marketing strategies within the industry have had significant effects as well, upon both the accessibility and – crucially – the price of foreign travel.

Package tours

The package (or inclusive) tour has been an especially important feature of post-1960 growth in international tourism although, as Towner (1996) makes clear, it is not an innovation. As early as the 1820s, travellers from London were able to purchase tours to France and Switzerland that were inclusive of all travel, food and accommodation, and the role of Thomas Cook as an innovator and developer of packaged tourism in the middle nineteenth century is well known (see, for example, Pimlott, 1947; Turner and Ash, 1975; Towner, 1996). However, in the post-1945 era, and particularly since the early 1960s, packaged forms of tourism have been central to the overall growth of the markets and the wider participation in international travel of less affluent groups. Initially, these impacts were largely confined to Europe, but the more recent development of long-haul forms of tourism has leant heavily upon packages as a basis to their growth.

The appeal of the package tour is very apparent. Williams (1996) argues that under the Fordist conditions of mass production and consumption that prevailed in the 1950s and early 1960s, a standardised product that was made affordable through the use of economies of scale was essential to meeting growing popular demand. He also suggests that alongside the low prices, the attractiveness of packages was reinforced through the comparative lack of knowledge and experience on the part of the tourist that increased their dependence upon the industry, as well as the practical advantage of purchasing an all-inclusive product in a single transaction.

The ability of the industry to meet these needs was the result of a rapid development from the mid-1950s of tour companies specialising in inclusive forms of travel. Initially these tended to focus on low-cost, mass markets visiting areas such as the Mediterranean, but by 1990 there were an estimated 800 companies in the UK offering foreign inclusive tours (Williams, 1996), including a significant number of small firms dealing in specialist tours to much more distant destinations. The market is, however, dominated by a few very large companies – such as Thomson, Air Tours and First Choice – which in 1995 accounted for 49 per cent of the British market for package tours (Mintel, 1998).

Most of the large firms have been created through mergers and this has been one of the chief pathways to creating travel companies with strong vertical linkages (for example, combining interests in accommodation, transportation and promotional or holiday retailing services) and with sufficient economic influence to negotiate favourable rates with local providers. The large firms have also been able to take advantage of their linkage patterns to extend their spatial 'reach' to bring successively more distant places into the package market. Hence, for example, Florida has become an affordable alternative to Spain or Portugal for British, German and Scandinavian package tourists. In these ways, large tour operators become enormously influential in shaping patterns of tourism, in both a spatial sense – through their capacity to foster public interest in new destinations – and a structural sense – through their ability to encourage particular segments of demand through the patterns of supply (Bote Gomez and Sinclair, 1996). For example, a study of the development of the Costa del Sol (Spain) by Barke and France (1996) shows how the development of low-cost package tourism not only initiated a dramatic expansion in the numbers of visitors, but also altered the social character of tourism to the region by encouraging disproportionately high levels of investment in cheaper category hotels.

Recent data on inclusive tours suggest that their convenience, reliability and relatively low cost continue to keep demand buoyant. Table 3.7 charts the growth in the inclusive tour market from Britain, Germany and Sweden between 1988 and 1995, although the data do conceal an important spatial

Table 3.7 Growth of inclusive tours markets in Britain, Germany and Sweden, 1988–96

Year	Tourists (millions)		
	Britain	Germany	Sweden
1988	12.6	17.6	1.3
1989	12.6	18.3	1.3
1990	11.4	17.8	1.2
1991	10.7	18.0	1.1
1992	12.6	20.2	1.3
1993	13.3	21.3	1.0
1994	15.2	22.7	1.2
1995	16.9	25.0	1.6
1996	15.7	27.0	1.7
% change, 1988–96	24.6	53.4	30.8

Source: Based on information from Bray (1996).

shift. This has seen the relative significance of package tours in short-haul journeys tending to diminish, whereas in long-haul markets, their importance has increased. For example, between 1986 and 1996, the share of the UK package market attributed to long-haul tourism rose from 7 per cent to nearly 16 per cent, with analysts predicting a further increase to a 25 per cent share by 2005 (Bray, 1996). The overall growth in foreign travel also conceals the fact that the share of foreign holiday markets that is taken by packaged tours has seen small but significant reductions as more tourists (perhaps having gained experience through the use of package tours) now travel independently. In 1986, inclusive tours accounted for 63 per cent of the UK foreign travel market, but by 1997 this had fallen to 56 per cent (Mintel, 1998).

Professionalism in the industry

The increased incidence of independent travel reflects the growing maturity in the industry and the confidence that tourists now place in the travel products that they are purchasing. Urry (1995) has provided a sociological interpretation of this process, emphasising the nature of tourism as a form of 'disembedding' in which social actions (such as leisure) are removed from the local contexts in which they normally reside and are recombined across larger spans of time and space. This, Urry (1995: 143) asserts, 'depends upon trust, people must have faith in institutions and processes of which they possess only limited knowledge'. The development of a professional travel industry in which people are prepared to place their trust is therefore seen as a key stage in overcoming perceived risks attached to travel to foreign lands.

Confidence is reinforced by the routine manner in which the industry is able to deliver its products. The ubiquity of the high street travel agency, the familiarity of the holiday brochure and the widespread promotion of foreign travel in the media, have all contributed to the demystification of international tourism. The simplicity of purchasing foreign holidays in 'one-stop' visits to agencies or by telephone and now Internet transactions, emphasises the ease of many forms of international travel and is an additional advantage.

Marketing strategies

Professionalism is linked to the increasing sophistication in the marketing of tourism services. A number of developments in the international travel industry have helped to raise public awareness of opportunities for foreign travel and, more importantly perhaps, driven down prices to levels that ordinary travellers can afford. These include:

- The increased use of market segmentation approaches. These aim to identify particular sectors within overall patterns of tourist demand – perhaps defined by age group, interests or motives – and allow for travel services to be tailored to particular needs and actively targeted at the segments concerned.
- Wider use of strategic alliances and consortia of providers. These may reduce costs (and hence prices) through activities such as joint advertising or the development of new products via reciprocal agreements. The creation of Europe-wide rail travel tickets are an example of the latter.
- Routine development of competitive pricing strategies. These will include bulk purchasing of hotel rooms and transport to construct low-cost packages; the use of discounting for early bookings and the use of APEX fares; reductions for off-peak/low season visits; and reduced prices for 'standby' tourists and people able to take up unsold holidays at the last minute. (See Horner and Swarbrook, 1996, for a comprehensive discussion of marketing strategies.)

The impact of technology

Part of the reason for the ease with which the modern travel industry can deliver its products is the application of new technology. Technology affects tourism in several distinct ways, but emphasis in this discussion will be placed upon two key contributions – in transportation, and in communications and information transfer.

The role of transport

In view of the increasing distances over which global forms of tourism now operate, the role of transport is crucial. Not only is it a means of getting to and around a destination but, as we have seen in Chapter 1, it is also an important part of the tourist experience in its own right (Page, 1999). Consequently, technological enhancements of transport systems have created several impacts upon tourism, but the most significant are arguably:

- the acceleration of services and the associated increase in the geographic range over which services are offered;
- the gradual reduction in the unit cost of foreign travel, particularly as the capacity of different systems – especially air travel – has increased.

All of the different transport systems contribute to the movement of international tourists, although their significance varies from place to place. In some accounts there is a tendency to present transport forms as successive in their impact, with a relatively simple progression from stagecoaches, via steamships and railways, to motor transport and the jet aircraft. However, as Voase (1995: 26) correctly emphasises, 'advances in the technology of transport have added to the repertoire of forms of transport available to the tourist, without eliminating any of the forms which have been in some ways "superceded"'. Even some archaic modes of travel (e.g. horse-riding) have re-emerged as integral elements within transport at tourist destinations, particular in niche markets such as adventure tourism.

This breadth of transport use is illustrated by the data in Table 3.8 which relate to the modes of transport used by European tourists and reveal how all the main forms continue to make a contribution. Because of the structure of the European market, the car is an important means of international travel, but as journey lengths increase to destinations outside Europe, the train and especially the aeroplane become much more significant.

To some extent, each of these forms of transport has been able to demonstrate the attributes of acceleration of travel, increase of range and relative reduction in costs that have been so important to the development of international tourism. Travel by car, for example, may have become slower within cities through increased congestion, but in inter-regional and international forms of road travel, the extension of motorway networks – especially between

Table 3.8 Variations in mode of transport used by European holidaymakers (%)

Destination	Car	Train	Plane	Boat	Coach
Own country	78	14	1	8	8
Other country	52	11	32	6	13
Non-EU country	53	15	29	17	18
Outside Europe	35	19	86	1	15

Note: Figures sum to more than 100% due to multiple responses.
Source: Page (1999).

northern and southern Europe – has enabled lengthy journeys to be undertaken quickly, in relative ease and comfort. After a period of decline brought on by poor performance and underinvestment, railway systems across western Europe are benefiting from new high-speed links such as the TGV services in France, ICE in Germany, AVE in Spain and Eurostar between London, northern France and Belgium (Pompl, 1993). The reduced journey times now enable trains to compete directly with air services on some routes, whilst discounted and promotional tourist fares make many foreign journeys by rail cheaper than domestic trips. Similarly, on short sea-crossings (such as between Britain, Ireland and northern Europe) accelerated services by hovercraft, jet-foil and high-speed, multi-hull boats have also been introduced.

However, it is in the field of aviation that impacts of technology are perhaps most clearly evidenced and most relevant to the growth of international travel. As Williams (1998: 54) notes, 'the expansion of international air traffic over the past twenty years has almost exactly matched the expansion in international tourism', and whilst the importance of air travel varies between different parts of the world, the development of key sectors such as European package tours and long-haul tourism to the Americas or East Asia and the Pacific is highly dependent upon aeroplanes.

According to Lyth and Dierikx (1994), in the formative phases of civil aviation between 1920 and 1940, tourists who flew were very few in number and, inevitably, drawn from elite groups. The situation was transformed initially by advances in aircraft construction – particularly in the USA during the Second World War – which made economic passenger transport by air a possibility. Then, after 1945, the availability of war-surplus aircraft at knock-down prices proved to be a stimulus to the creation of the first air tour companies to take advantage of the newly available technology – Horizon (1953) and Skytours (1954). By 1960, the introduction of the Boeing 707 jet airliner had signalled the general demise of piston-driven planes on scheduled services and this provided a ready supply of cheap planes just as the expansion in European package tours was beginning to accelerate. By the late 1960s, smaller jet planes such as the Boeing 737 were being added to the fleets of operators, cutting journey times between northern Europe and the Mediterranean by up to 40 per cent. The impact of these developments on air charters was dramatic, with the British market expanding from 537,000 passengers in 1963 to over 7.7 million in 1972 (Figure 3.5).

In the early 1970s, the introduction of the Boeing 747 'Jumbo' airliners (and similar wide-bodied, long-range jets) initiated yet another round of development, by extending the range of non-stop flights to over 10 hours and increasing payloads to over 400 passengers per flight. The economies of scale that the new technology provided, when allied with structural changes in the aviation industry that have seen gradual removal of monopolistic powers, deregulation, and wider use of 'open skies' systems (Page, 1999), have therefore contributed directly to the development of the long-haul, global tourist markets in areas such the Americas, Africa, East Asia and Australia that were outlined in the first part of this chapter.

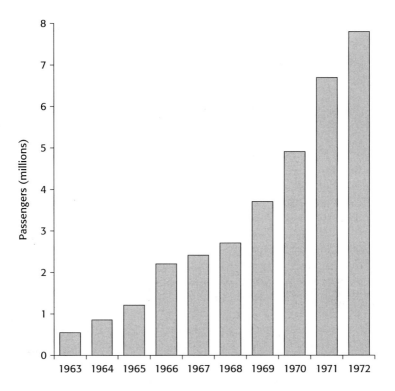

Figure 3.5 Expansion of the British air charter market, 1963–72
Source: Based on information from Lyth and Dierikx (1994).

Information technology

Aviation companies have also been in the forefront of the second key technological development in international tourism, the introduction of information technology. In an industry where information is the lifeblood, the application of computer-based systems has a clear potential to revolutionise operations and the relationships between consumers and providers, by increasing both the availability of information and the speed and convenience with which it is made available (Buhalis, 1998; Pollock, 1995).

Information technology has penetrated tourism at a number of different levels. For example, it is now routine for airline companies, hotel groups, transport companies and car hire firms to run internal database systems to handle bookings or provide information to individual clients (Truitt *et al.*, 1991). Increasingly, computer systems are also being used to speed routine procedures within tourist journeys, such as airport check-ins (Knowles and Garland, 1994), whilst technology such as CD-ROM and DVD are anticipated to make significant contributions to the work of travel agents in presenting destination information to intending foreign travellers (Archdale, 1993). But the greatest scope for influential change lies in the development of globalised systems of information distribution.

In reviewing the growth of such systems, Buhalis (1998) proposes a three-stage summary of developments:

1. Computer reservation systems (in the 1970s).
2. Global distribution systems (in the 1980s).
3. Internet and the World Wide Web (in the 1990s).

Computer reservation systems (CRS) were primarily developed by airline companies as a means for centralised control over the sale and distribution of airline seats. The ability of these systems to provide instant information on seat availability and pricing, and to make confirmed bookings secured by credit cards, greatly simplified a key tourist transaction. Knowles and Garland (1994) state that the airline CRS were initially geared towards business travellers, but as these markets became saturated, the scope widened to include leisure tourists. Unsurprisingly, the benefits of the approach were quickly recognised by providers other than airlines, and major hotel chains and car rental groups also developed CRS of their own (Go, 1992).

Global distribution systems (GDS) are an elaboration of CRS, but expanded to provide a much broader range of information, both spatially and through vertical and horizontal linkages – as a form of 'electronic travel supermarket' (Buhalis, 1998). Because of the additional complexities in extended information systems, GDS have tended to be developed by separate firms, albeit with companies such as airlines as major shareholders. Currently four systems dominate the world market – Galileo, Amadeus, Sabre and Worldspan – offering information, direct booking and ticketing facilities on a growing range of travel services, accommodation and entertainment across the globe. In 1996, for example, the Amadeus GDS was linked to over 106,000 travel agencies worldwide, displaying online information on 432 airlines, 29,000 hotels and 55 car rental companies (Buhalis, 1998).

GDS still require tourists to contact agencies who act as 'gatekeepers' to the information at their disposal, but the development of Internet services offers a quite different approach with some particular advantages to the tourist. First, the World Wide Web, as a multimedia facility, is able to combine, text, pictures, graphics, video and sound to present tourists with information in novel, attractive and influential ways. As more people subscribe to Internet connections to their home, this increases the ease and convenience of browsing for tourism-related information. Second, and perhaps more importantly, the Web empowers people by releasing to them information and opportunities that were previously unavailable, or at least, harder to retrieve and utilise. Once empowered, simple telephone, e-mail or online transactions between consumer and supplier offer rapid confirmation of travel purchases, without the consumer ever visiting conventional travel agencies or brokers. Thus, although the Internet is still in its infancy, it will clearly exert a fundamental impact on the way travel is marketed, distributed, sold and delivered (Pollock, 1995).

The impact of economic and political convergence

The importance of enhanced information systems is that they simplify the business of travel and in so doing, encourage participation. The same is generally true of the impact of convergence and associated changes in the stability of global political and economic systems. The potential impacts of such changes are especially well illustrated in Europe, and particularly within the EU. Here, tourism has been stimulated not only by the political stability and increased affluence that has broadly followed in the wake of the Treaty of Rome, but also by a range of more specific changes.

Urry (1995) draws attention to several dimensions of change. First, he emphasises the internationalisation of tourism enterprise. The creation of the EU as a trading bloc in which protective barriers to trade and commercial operations have been largely removed, directly enables the development of travel companies with international portfolios of investment and ownership (see Davidson, 1992). As a consequence, provision of tourist facilities is extended; the development of packaged, integrated tourist products is made easier; and greater competition matched with the economies of scale helps to keep prices competitive. Deregulation of key sectors – in particular, airline companies – has also helped to reduce costs by opening more routes to competition.

Second, changes in the way travel is regulated and supported have made international tourism easier. At a global scale, fewer countries now require visas as a prerequisite to entry, whilst within the EU, border controls have been removed from large swathes of territory following the Schengen Treaty of 1995. (Belgium, France, Netherlands, Luxembourg, Germany, Portugal and Spain all permit unregulated movement of tourists between their territories.) Associated developments of Europe-wide institutions have also been beneficial in, for example, the reciprocal provision of health care for visitors.

Third, financial transactions have been simplified. The development of globally recognised credit cards (e.g. Visa, Mastercard and American Express) have reduced the need for tourists to exchange (or even carry) significant amounts of foreign currency, thereby saving on the costs of foreign currency transactions. More significantly, the acceptance across large parts of the EU of the euro as a single currency, will simplify transactions, reduce the costs of both the production and consumption of tourist products and services, and remove the uncertainties associated with short-term fluctuations in the exchange rates between different currencies.

Lifestyle changes

The development of a sophisticated travel industry; the impacts of innovation in transport and information transfer; and the growing ease of travel in an international community are all relevant factors, but it is arguably the changes in lifestyle that are the key to the development of foreign tourism. Without the public appetite for foreign travel – without *demand* – none of the factors discussed so far is capable, in isolation, of creating tourist patterns. In particular, changing levels of affluence, the fashionability of foreign travel and the developing competence and sophistication of contemporary tourists, have all helped to widen the base for participation in foreign travel and extend its spatial range. Consequently, as Mintel (1996: 7) observe, '[foreign] holidays are increasingly entwined with leisure lifestyles'.

Although the relative costs of foreign travel have been driven downwards throughout the post-1945 period, the basic relationship between affluence and travel is still relevant. In Britain, as in most Western nations, recent trends have undoubtedly favoured the growth of holidays abroad as affluence levels have generally risen. Data in Table 3.9 show that although social polarisation has been one feature of structural change in the UK since 1980 (with more people moving into the lowest E category, especially through retirement), the increased proportion of people in the upper ABC1 social categories (and the rapidity with which these groups have expanded) identifies a growing potential market for foreign travel.

One of the outcomes of these shifts has been an increased incidence of multiple holidays, in which the proportion of the population able to take more than one holiday has risen and within which, the significance of a foreign trip as a fulcrum of holiday patterns has become more pronounced. Figure 3.6 shows how the proportion of British tourists taking more than one holiday per year has changed between 1972 and 1996, whilst Table 3.10 reveals how recent changes within multiple holiday patterns have favoured foreign travel. According to these data, amongst Britons who took a holiday in 1997, an estimated 55 per cent would be taking their main holiday abroad (Mintel, 1997d).

As more sections of society take up foreign travel, so the habit is likely to become fashionable. In many communities, the ability to take a foreign holiday has emerged as a mark of status which, through time, fosters the progressive inclusion of more participants. The media, in particular, actively promote and promulgate images of foreign travel as an enticing yet accessible component of

Table 3.9 Changes in the socio-economic structure of Great Britain, 1980–2000

Socio-economic class	1980		1990		2000		% change
	millions	%	millions	%	millions	%	
AB	7.40	17	8.40	18	11.36	24	+53.5
C1	9.90	22	10.90	23	13.75	29	+38.9
C2	14.24	32	13.48	29	10.32	21	−27.5
D	9.23	21	8.16	18	7.28	15	−21.1
E	3.75	8	5.73	12	5.31	11	+41.7

Source: Mintel (1997d).

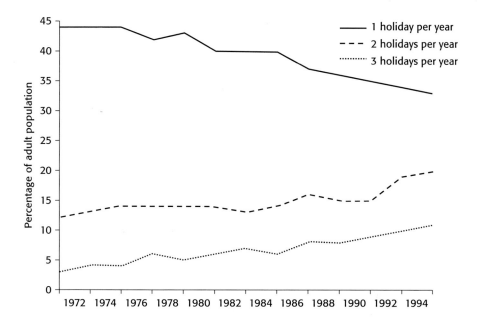

Figure 3.6 Patterns of multiple holiday-taking by UK residents, 1972–96
Sources: Based on information from BTA (1995); Mintel (1997d).

Table 3.10 Recent changes in location of holidays taken by British tourists

Pattern	% of those who take a holiday		
	1991	1995	1997
All holidays taken abroad	18.0	16.9	20.9
Main holiday abroad, second holiday in UK	30.2	38.5	34.5
All holidays in Britain	45.8	37.3	38.3
Other combination	6.0	7.3	6.2

Source: Adapted from Mintel (1997d).

contemporary lifestyle, raising awareness and instilling in populations positive desires to travel and to see previously imaged sites (and sights) for themselves.

Hence, the experience of travel (and the greater levels of empathy towards the notion of travel) enhances the cosmopolitan dimensions of society which, in turn, make people more susceptible to international forms of tourism. Urry (1995: 167) defines cosmopolitanism as 'an intellectual and aesthetic stance of openness towards divergent experiences from different national cultures' and argues that one of the general effects of international tourism is to generate wider incidence of what he terms 'aesthetic cosmopolitan' attitudes. The dynamics of aesthetic cosmopolitanism are summarised in Table 3.11, but it is

Table 3.11 Dimensions of 'aesthetic cosmopolitanism'

- Extensive patterns of real and simulated mobility in which one assumes the right to travel anywhere and to consume, at least initially, all environments
- A curiosity about places, peoples and cultures and at least a rudimentary ability to map such places and cultures historically, geographically and anthropologically
- An openness to other peoples and cultures and a willingness/ability to appreciate some elements of the language/culture of the place that one is visiting
- A willingness to take risks by virtue of moving outside familiar tourist environments or situations
- An ability to locate one's own society and its culture in terms of a wide-ranging historical and geographical knowledge, to have some ability to reflect upon and judge aesthetically between different natures, places and societies
- A semiotic skill to be able to interpret tourist 'signs', to see what they are meant to represent and to know when they are partly ironic

Source: Adapted from Urry (1995: 167).

important to appreciate that not only are cosmopolitan attitudes a consequence of tourist experience, but also, they are a stimulus to further exploration, particularly as society in general – and its more mobile groups in particular – adopts a more reflexive view of the value of different physical and social environments.

It should be recognised, too, that the cosmopolitan tourist is essentially a competent tourist and as personal experience of travel becomes routinised, so the competence that comes with experience influences not only the growth of mass tourism sectors, but also permits tourists to indulge new leisure tastes and exercise greater degrees of choice. This is evident in several directions, including:

- wider participation in independent forms of travel;
- increased commitment to foreign travel, for example, through the purchase of timeshares or second homes;
- the growth of special interest forms of tourism and the creation of new market segments – such as adventure tourism or activity holidays based around popular sports and recreations (see Mintel, 1997c).

Independent travel abroad by the British, for example, has risen from 3.2 million visits in 1975 to 13 million in 1995 (BTA, 1995; Mintel, 1996), although the market share taken by independent travel has grown more moderately. Similarly, self-catering holidays taken abroad (which are another indicator of tourist competence) have risen amongst British tourists from just 350,000 in 1975, to over 6.8 million in 1993 (BTA, 1995). Part of this growth – albeit a small element – is attributable to the rise in second-home ownership and timeshare properties. Mintel (1997b) estimate that around 35,000 Britons currently own second homes abroad – especially in the accessible regions of northern France and Belgium – whilst timeshare ownership had risen to

337,000 in 1997 and covers a wider geographic area that includes southern France, Greece, Spain, Portugal and – as a more recent development – Florida.

Part of lifestyle change in areas such as the UK, Europe and North America has seen a growing public interest in healthy living, the impact of which upon recreational patterns has been to foster growth in participation in activities such as walking, swimming, cycling and golf, amongst a widening range of sports and active recreations. The absorption of these activities into tourist patterns has been selective, partly because for many people the motivation for tourism lies in escape from routine and the chance to indulge what Graburn (1983) characterises as behavioural 'inversions'. Thus active, healthy lifestyles are often temporarily substituted by relaxation and exceptional patterns of consumption. But as Mintel (1997c) note, whilst many forms of tourism may be typified by inactivity, active forms of holidaymaking in which popular recreations form both a rationale and a focus for the visit, are becoming more established. In the UK, some 14 million activity holidays are taken each year, 20 per cent of these being taken abroad. Moreover, whilst levels of domestic activity holidays have remained comparatively stable, the smaller foreign holiday market grew by some 14 per cent between 1994 and 1997 (Mintel, 1997c). Foreign activity holidays centre upon a range of popular recreations (including walking, cycling, golf and sailing or boating), but it is perhaps the emergence of skiing holidays that exemplifies best the synthesis of recreation, tourism and lifestyle that this section is discussing.

Case study: Skiing in Europe

Skiing – as a fashionable winter sport and holiday activity – is a quintessential expression of lifestyle through recreation and tourism. Originating as a socially exclusive activity amongst upper-class European tourists before the First World War, it retains, to the present, a glamorous and rather privileged image. As with most sectors of tourism, however, a gradual process of democratisation has prompted a broadening of participation and an extension of provision. Thus, by the 1920s, well-defined winter resort areas had emerged in the Alpine regions of France and Switzerland. At this time Swiss resorts such as St Moritz and Davos each afforded nearly 6,000 bedspaces, whilst ski resort development around Chamonix in France not only matched many of the Swiss resorts in scale (Towner, 1996), but was also instrumental in shaping a new pattern of higher-altitude development that eventually became influential in later phases of development.

The tourist skiing market prior to the 1960s was focused upon an essentially small and exclusive clientele (although as a recreational interest amongst resident populations in areas where skiing is possible, the activity had formed an integral element in local leisure lifestyles for an extended period). The development of skiing as a popular winter tourist activity emerges only with the growth of cheaper (although not cheap) package holidays after 1970. Estimates suggest that the 1970s were a particularly important phase for the development of the European skiing market, with annual growth in the order of 7 per cent. During

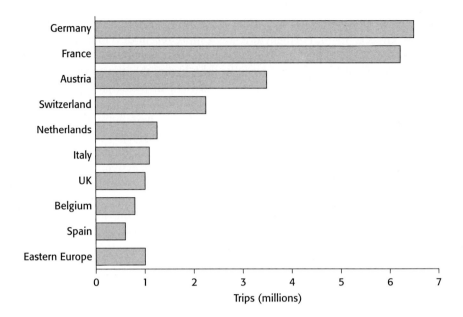

Figure 3.7 The European ski holiday market, 1998
Source: Based on information from Smith and Jenner (1999).

the 1980s that rate of development slowed to around 3 per cent per annum, although in markets such as the UK, growth was more pronounced as the activity became an attractive component in the lifestyle of young, wealthy professionals (Mintel, 2000a).

By 1998 the European ski holiday market (which combines foreign as well as domestic trips) stood at 24 million visits annually – see Figure 3.7. Within Europe, Austria accounted for 44 per cent of the market, followed by Switzerland (16 per cent), France and Italy (both 13 per cent) (Smith and Jenner, 1999). The popularity of Austria (which is built around an ideal combination of excellent skiing conditions with traditional landscapes, cultural images and entertainment) is, however, sustained primarily by visits from its German neighbour which generates just over half of all overnight stays in ski resorts (Smith and Jenner, 1999). Amongst some of its other traditional markets, though, Austria has become much less attractive. A loss of markets has occurred for several reasons, but unfavourable foreign exchange rates and a number of poor snow seasons have been particularly influential in encouraging some tourists to look elsewhere. An analysis of the British ski market, for example, reveals a marked decline in the attraction of skiing in Austria and an increased preference for French and North American ski resorts. Between 1988 and 1998, the proportion of the UK ski market that visited Austria fell from 50 per cent to just 12 per cent, being replaced by France (33 per cent) and North America (20 per cent) as the favoured destinations (Smith and Jenner, 1999; Mintel, 2000a).

The ability of British skiers to refocus their attentions onto a long-haul destination (North America) emphasises the fact that the ski market remains the domain of primarily affluent people for whom a skiing holiday is a key component in a fashion-conscious lifestyle. In the UK, 82 per cent of skiers are in the upper ABC1 social groups; across Europe as a whole, nearly 70 per cent are drawn from upper or upper-middle groups; whilst in the USA, 70 per cent of skiers came from households with an annual income of at least US$50,000 per annum (Smith and Jenner, 1999). Skiing is also primarily a young person's activity. The vast majority of downhill skiers are under 45 years old, although in less taxing forms of skiing (such as cross-country), older participants are more numerous.

The age-selective nature of participation is a consequence of the physical demands that skiing can place on participants. The dominance of the wealthier social groups demonstrates, in part, the costs of equipment, of travel, of lift passes and of accommodation in what are often up-market resorts. But there is evidence, too, that the image attached to skiing deters participants. Smith and Jenner (1999) observe that many people in skilled occupations could afford the costs of participation, but choose not to on the basis of the perceived image that surrounds the activity.

Although the European ski market is still a substantial one, through the 1990s the mature destinations of France, Austria and Switzerland have largely stagnated, although growth continues in parts of eastern Europe (Smith and Jenner, 1999). Stagnation has been attributed to a number of factors, including:

- short-term economic problems associated with recession at the beginning of the 1990s;
- an ageing population in the main tourism-generating countries;
- adverse publicity associated with skiing accidents and fatalities of skiers in avalanches;
- wider incidence of poor snow conditions attributed to global warming. This has been blamed for a contraction in the length of the season, with fewer visitors risking skiing holidays in December and January, when snow is less certain;
- a redistribution of skiers to newer, competing destinations – such as the USA and Canada (Smith and Jenner, 1999).

Thus, although new markets (especially in eastern Europe) are being developed and new snow sports (such as snowboarding) are finding popularity with some tourist groups, analysts paint an uncertain picture of the period ahead for skiing. Interestingly, one of the factors cited by Mintel (2000a) in accounting for present uncertainties is the wealth of alternative activities and attractions around which tourist trips may be centred – reinforcing, perhaps, the intimate relationships between wider recreational lifestyles and tourism patterns that this book is addressing.

Independence in foreign travel and the growth of a range of activity holidays are ultimately a consequence of the wider exercise of consumer choice which, it has been argued, characterises the leisure lifestyles of post-industrial societies (Poon, 1989). This has helped to stimulate the development of new market segments and many forms of special interest tourism. This is shown in:

- the growing number of destinations that tourists are now able to visit;
- the diverse nature of attractions that are visited within destination areas;
- the temporal extension of travel into periods of the year (especially winter) that were previously considered exceptional.

This flexible, often eclectic nature of contemporary tourism and its wider integration into personal lifestyles, has become one of its defining features and has formed a central element in the changing relationships between tourism, recreation and leisure.

Convergence between recreation and foreign travel

This brings us back to the theme of convergence between recreation and tourism and the diminishing differentiation of the two within contemporary lifestyles. The chapter opened by suggesting that whilst foreign travel bears many distinctive features that sets this form of tourism apart, a closer relationship between foreign travel and other forms of recreation and leisure is emerging. From the preceding discussions, this is evident in several ways.

First, the discussion above argues that whilst there have been a host of enabling factors shaping the recent development of international tourism, it is the changes in personal lifestyle that are fundamental in creating and shaping demand. These changes reflect the wider shifts within post-industrial society across the westernised world in placing a new emphasis upon processes of consumption (rather than production) in shaping the way we live. As the discussion in Chapter 1 has shown, recreational tastes and preferences have become progressively more central to defining our own identities and our expression of identity through styles of living. Hence, at the broadest of levels, the choices that we make as tourists of where we visit and what we do, often reflect (by extension) recreational preferences that shape routine leisure. These links are then progressively reinforced by the travel industry through the growing use of segmentation in structuring the market and in promoting its tourism products, a process that builds upon an understanding of our preferences as tourists that is informed by our wider lifestyles and expressed leisure choices.

This becomes evident, second, in tourist behaviours and activity patterns. Whilst we may spend significant amounts of time and money in decamping to foreign places, the way we fill our time at destinations is typically based around familiar recreations and leisure activities. We swim, read, sunbathe, drink, eat, dance, play sports, shop and – language barriers permitting – watch television, much as we might do in our leisure time at home. The context may be different, but many of the activities are not. Increasingly, too, we not only build favourite recreations and pastimes into our foreign holidays, we structure foreign travel around them, as the discussion of the growth of activity holidays and the case study of skiing illustrate. This dependence upon the familiar for at least a part of the tourist experience is not, of course, a new idea. Boniface and Fowler (1993: 13), for example, echo the much older arguments of Boorstin (1964) when they comment that 'deep down, a lot of people travel to arrive where they came from'. But, in the context of the argument in this book, the point needs re-emphasising.

Third, we should note the temporal and, especially, the spatial convergence of recreation and tourism that is evident in some of the changes outlined above. As travel has become faster (and in relative terms, cheaper) so the incidence of day or short-break travel to foreign destinations has become more apparent, especially within Europe. International day trips, in particular, confuse the distinctions between recreational trips and tourism, when the latter is defined (as by tradition it has been) as embracing nights spent away from home. They confound our instinctive expectations that a foreign trip will be clearly identifiable as a form of tourism and make distinctions harder to draw.

Not only are temporal distinctions less clear, but spatial differences are too, as many day visits and short excursions typically bring foreign tourists into the same spaces that are occupied by local recreationalists. This is well illustrated in sectors of the market such as city-break tourism (where shops, entertainment, museums, galleries, restaurants and public open spaces attract local people, recreational visitors and tourists, without distinction) and in the growing appeal of theme parks and commercial attractions to recreational visitors and tourists alike.

In a different way, spatial convergence may also be evident in the way in which landscapes of international resorts are often remodelled to suit the tastes and preferences of core tourist groups. Earlier in the chapter the problems of Spanish package holiday resorts were briefly mentioned. But alongside the rising incidence of overdevelopment, commercialisation, pollution and the range of social problems that has affected resorts such as Benidorm and Torremolinos, a more fundamental loss of Spanish identity has been evident. The English-style pubs, fish and chip shops and nightclubs that dominate the tourist areas of these resorts impose an identity that owes little to indigenous traditions and much to the recreational habits of the British package tourists on whom the resorts now depend.

Finally, we should reflect again on the nature of tourist experience that was examined in Chapter 1. This model emphasises the point that tourism is not just about the experiences at the destination, or the journey there and back – important though they may be. The tourist experience has a significant phase of planning and anticipation and, perhaps most important of all, a place in our memories. These phases are, by definition, home-based – occupying leisure time in both the planning and retrospective phases, perhaps involving recreational interests such as reading (in planning a trip) and photography or home videos (in acts of recall), and often set within the social context of family leisure. Thus whilst the act of travelling abroad may be delineated as tourism, it resides within a wider context that is shaped by our use of leisure and our recreational interests.

Questions

1. What have been the key changes in the spatial patterns of international tourism since 1990, and why have these changes occurred?
2. Outline the primary impacts that the development of package holidays has had upon the expansion of international tourism.
3. How has technological change assisted the recent development of international tourism?
4. Examine the connections between international tourism and lifestyle.
5. In what ways may the experience of international tourism and local recreation be seen to be convergent?

Further reading

Historic patterns of development in international tourism have been given comprehensive coverage in Turner, L. and Ash, J. (1975) *The Golden Hordes: International travel and the pleasure periphery*, London: Constable; and more recently in Towner, J. (1996) *An Historical Geography of Recreation and Tourism in the Western World, 1540–1940*, Chichester: John Wiley.

For an interesting collection of essays reviewing tourism in a major international destination, see Barke, M., Towner, J. and Newton, M.T. (eds) (1996) *Tourism in Spain: Critical issues*, Wallingford: CAB International.

The role of transportation in international tourism has been fully covered in Page, S.J. (1999) *Transport and Tourism*, Harlow: Addison-Wesley Longman.

An excellent review of the impacts of information technology is provided by Buhalis, D. (1998) 'Strategic use of information technologies in the tourism industry', *Tourism Management*, **19** (5): 409–21.

Regular reviews of key market sectors (including excellent statistical coverage) are provided by the market research group Mintel (UK) Ltd in their publication *Leisure Intelligence*. For statistical coverage of tourism at a global scale, the World Tourism Organization produces an annual *Yearbook of Tourism Statistics*. Data may also be accessed via their website – http://www.world-tourism.org, although users are currently obliged to subscribe to receive the full service.

Tourism and recreation in urban places

CHAPTER 4

Introduction

Urban places occupy a pivotal position in the development and practice of recreation and tourism – a simple truth that stands scrutiny even if academic recognition of the fact has sometimes been inconsistent and uneven. In discussing recreation in urban places, Williams (1995) supports this view by noting three related facts. First, in all developed (and many developing) nations, the majority of the population live in towns and cities – in countries such as the USA and the UK, overwhelmingly so. Second, the localised nature of many routine recreational events ensures that most leisure time is spent within the same urban environments in which the majority reside. Third, the popular recreations that fill our leisure time tend to be either home-based or are accommodated in built provision (such as leisure centres, commercial entertainment, bars, restaurants, shops and cultural facilities) that also are naturally concentrated in urban places.

The links between tourism and urbanism are similarly intimate. As Page (1995) makes clear, much of the demand for domestic and international tourism is generated within urban populations and although, historically, tourism patterns were often shaped by an evident desire to escape the urban environments of the major industrial cities and conurbations (Tuppen, 1996), many of the destinations to which people travelled – especially seaside resorts – were still urban places, albeit of differing character. Moreover, in an intriguing reversal of patterns, major cities are now emerging as key tourist attractions within post-industrial leisure patterns. Several authors (including Burtenshaw et al., 1981, 1991; Law, 1996) remind us that in simple numerical terms, major cities such as London, Paris, Rome, Amsterdam and New York easily outrank the most popular leisure resorts in the levels of visiting that they receive. In 1994, for example, London received an estimated 424 million domestic day visits (CRN, 1996) and over 90 million tourist nights were spent by domestic and international visitors in the city (Bull and Church, 1996).

However, although the significance of urban recreation and tourism is becoming more evident, the roles that urban places play as centres of leisure activity and the relationships between recreation and tourism in towns and cities are far from clear. The lack of clarity arises for several reasons.

First, there are significant problems in identifying a distinctive resource base for urban tourism, given that many of the attractions that draw tourists (e.g. parks, historic buildings, shops, restaurants, entertainment and cultural facilities) are also used extensively by local people. In the West End of London, for example, theatre audiences typically comprise around 40 per cent local residents, with the remainder divided fairly evenly between domestic and international tourists (see Law, 1993). Tourism thus forms part of a range of urban functions but it is often difficult to isolate and measure the particular dimensions of tourism as distinct from local recreational and non-recreational activity (Law, 1996).

From this it follows, second, that spatial patterns and, especially, distinctions between recreation and tourist spaces are similarly imprecise. Whilst most urban places have well-defined tourist action spaces (and equally, large parts of the city in which visitors seldom venture), tourist space is also used by local residents for recreation, creating confusing patterns of spatial coincidence. Hall and Page (1999: 139) write that 'within cities . . . the line between tourism and recreation blurs to the extent that at times one is indistinguishable from the other, with tourists and recreationalists using the same facilities, resources and environments'.

Third, tourism in cities gains added complexity from the range of reasons for which people may visit. These include recreational and holiday purposes, visiting friends and relations, business purposes (including conference attendance), shopping trips and educational visits. Many trips are also multifunctional, combining a range of objectives that take advantage of the diversity of opportunity that urban places generally provide (Shaw and Williams, 1994).

Lastly, the dynamic nature of deindustrialised urban places ensures that new opportunities for recreation and, especially tourism, are being repeatedly created and recreated within ongoing processes of urban redevelopment. Hence, both spatial and behavioural patterns of tourism and recreation may be subject to frequent readjustment and change, some of which may be unpredictable but which appear to be promoting a progressive overlap and coincidence of urban tourism and local recreation.

In reflecting these uncertainties and confusions, Law (1996) echoes a number of authors who have questioned the extent to which the term 'urban tourism' constitutes a meaningful description and asks whether, in practice, it is a chaotic concept that describes a range of phenomena that are only loosely related. The question is perhaps rhetorical, but it underlines the uncertainty that surrounds the investigation of recreation and tourism in the urban environment.

This chapter aims to explore the dominant patterns of urban recreation and tourism as well as the evolving relationship between the two sectors of leisure. The next section examines and explains some basic characteristics of urban recreation and tourism. This is then linked to a discussion of key processes of change in the use of urban recreational facilities and finally, to a consideration of the emergence of new urban leisure spaces and practices in which recreation and tourism are frequently combined. In this latter section, the theme of convergence between urban recreation and tourism is exemplified through a discussion of two areas of interest that have developed significant levels of popular appeal: leisure shopping and sport as a spectacle.

Urban recreation and tourism: place, space and activity

Tourist space and activity

It is clearly inappropriate to presume that the recent recognition of urban tourism within academic literature signifies a similarly abbreviated history of tourism to urban places. Macdonald (2000) summarises a chronology of leisure visiting which demonstrates that whilst tourist use of urban places has not always been consistent through time, it nevertheless ranges over an extended period. Thus Macdonald catalogues:

- patterns of pleasure visiting to urban sites in the ancient civilisations of Greece and Rome;
- the widespread incidence of religious 'tourism' – as pilgrimages – across medieval Europe and the Middle East;
- the emergence of an elite and aestheticised form of urban tourism within the European Grand Tour in the seventeenth and eighteenth centuries;
- the reinvention of medicinal spas as centres of urban leisure in the eighteenth and early nineteenth centuries;
- the development of urban spectacles as tourist attractions in the later nineteenth and early twentieth centuries – including major exhibitions in cities such as London and Paris, and later, the Olympic Games;
- the integration of urban tourism into city regeneration and development towards the end of the twentieth century.

However, there is no doubt that since, perhaps, 1970, the incidence of urban tourism has grown significantly in scale and has affected a much wider cross-section of urban destinations than previously. Within western Europe, for example, the international market for urban tourism in 1995 was estimated at 124 million trips per year, of which perhaps one-third were holiday visits. Although many of these trips are of a characteristically short duration with, for example, in the holiday sector short city breaks accounting for two-thirds of

Table 4.1 European city holiday trips – major destinations, 1995

Destination	Total no. of trips ('000)	Market share (%)
France	4,700	15.2
Germany	3,400	11.0
United Kingdom	3,160	10.2
Italy	2,300	7.4
Austria	1,970	6.4
Czech Republic	1,500	4.8
Netherlands	1,370	4.4
Belgium/Luxembourg	1,090	3.5
Spain	940	3.0
Hungary	800	2.6
Denmark	550	1.8
Sweden	540	1.7
Portugal	470	1.5
Poland	450	1.5

Source: Adapted from Cockerell (1997).

the overall market, the total trade is still a substantial one. At a national level, the European market is dominated by France – largely thanks to the appeal of Paris – together with Germany, the UK, Italy and Austria. However, as Table 4.1 illustrates, urban tourism is spread widely and quite evenly over the majority of European states, and although traditional urban destinations such as Rome, London, Amsterdam, Paris, Venice and Vienna continue to exert a strong appeal, newer urban tourism destinations are emerging. These include cities such as Athens, Barcelona, Madrid, Prague and Warsaw, and this points to the way in which growth in the overall market is also promoting a spatially broader range of urban tourist places (Cockerell, 1997).

The factors that have stimulated the rise of urban tourism are grounded, in part, within the wider availability of leisure time and holiday periods, the increased levels of affluence and the greater levels of mobility that have affected tourism and recreation patterns generally (see Chapter 1). Changing patterns in the use of holiday periods – especially the increased incidence of multiple holiday-taking – have been influential as well. In particular, the rising popularity of off-season, short-break forms of tourism has clearly favoured urban locations.

But there are also more specific factors at work. The diverse nature of urban tourism – with its range of motives and purposes – has been identified briefly in the introduction, and this points towards a broader set of explanatory factors. Law (1996), for example, argues that the incidence of visiting friends and relations is a product of changing patterns of mobility in employment and retirement that has seen a progressive, spatial extension of kinship and friendship networks. Similarly, the time–space compression that authors such as Harvey (1989) argue is emblematic of post-industrial life, affects activities and events that draw visitors from vast catchments – especially for major urban spectacles. Rising levels of education and the power of global media to disseminate ideas and images also help to enhance the desire to see major urban sites, whilst the acceleration of transport services and the relative reduction in the cost of travel make it increasingly possible to do so.

Page (1995) extends this line of argument by identifying a range of structural factors that have helped to encourage urban tourism. These include:

- the significance of many urban places as nodal points within internal transport networks or as gateways for international travellers. This ensures a high level of incidental contact between tourists and cities, even if primary destinations lie elsewhere;
- the concentration of commercial, financial, industrial and producer services in urban places encourages high levels of employment- or business-related tourism;
- similarly, the concentrations of cultural, artistic, entertainment, sporting and recreational attractions (many of which owe their origin to local demand as much as tourism) attract the holiday visitor.

Important, too, is the extent to which the range and scale of urban attractions have been extended as post-industrial patterns become more firmly established in towns and cities. The scope for tourism to act as a catalyst to development within urban regeneration programmes was first demonstrated in the USA in the 1970s at locations such as the Inner Harbour in Baltimore (see Blank, 1996), but the practice has been widely imitated in European industrial cities seeking to reinvent themselves as post-industrial places. This process has entailed not only the conversion of redundant urban zones into areas of retailing, residence and leisure, but also conspicuous investment in new, prestigious forms of provision (centred upon cultural facilities, hotel and convention complexes, indoor arenas and entertainment) as active components in urban place promotion (Gold and Ward, 1994). Sites of world renown – such as the Sydney Opera House – catch the eye, but regeneration programmes with a more localised purpose are far more numerous and, arguably, far more influential. The development of Birmingham's Centenary Square (Figures 4.1 and 4.2) illustrates one example of a regeneration programme with significant regional and national impacts.

These processes of redevelopment both reflect, and influence, changing public tastes and preferences, particularly with their emphasis upon new forms of attraction such as leisure shopping and urban heritage sites. The extension of choice that confronts the modern urban visitor has prompted Law (1996: 7) to describe major cities as 'the ultimate post-Fordist, post-modern destination' – replete with alternative attractions and spaces, and brimming with leisure opportunities.

The nature of the attraction of urban places to tourists – the urban tourism product, in marketing terms – has been usefully conceptualised by

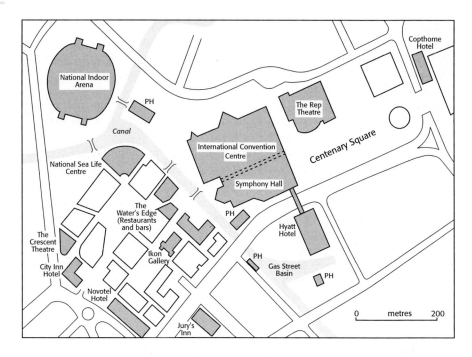

Figure 4.1 Tourist and recreational developments around Centenary Square, Birmingham

Jansen-Verbeke (1986). She argues that the essential elements that comprise the basis for urban tourism may be organised under four headings. The primary elements are made up of what Jansen-Verbeke terms 'the activity place' and 'the leisure setting'. The former comprises a range of place-specific attractions and facilities that help to define the more tangible dimensions of the urban tourist product (such as cultural and entertainment facilities, or events), whilst the latter attempts to delineate the broader context in which tourism is placed. Hence, the leisure setting includes different physical situations in which tourism might develop (for example, parkland or waterfront areas), as well as acknowledging the significance of the social and cultural context (such as local custom and patterns of life). These are often critical in defining the tourist's sense of place or its place identity – images that are then reused (and hence reinforced) through the marketing of cities as tourist attractions.

Supporting the primary elements are several 'secondary' and tertiary elements (or 'conditional elements' as they are labelled by Jansen-Verbeke). The secondary elements include accommodation, catering and retail facilities, whilst the conditional elements contain basic infrastructure such as parking, information, signage and guides.

The original model has been widely replicated within the literature on urban tourism but has also been quite widely criticised, principally because the diversity of motives that underpin tourist visits makes the separation of primary and

Figure 4.2 Part of Birmingham's Centenary Square–Brindley Plaza redevelopment, with Symphony Hall and the International Convention Centre (right); the National Indoor Arena (background) and the Water's Edge complex of shops and restaurants (left)

secondary elements – as suggested by Jansen-Verbeke – problematic. That is, what may be a secondary element for one visitor (for example, retailing) may be a primary element for someone else. Figure 4.3 therefore represents a reworking – and a simplification – of Jansen-Verbeke's original idea that attempts to draw a sharper distinction between primary and supporting elements within the supply of urban tourism. In reading the diagram, however, it is important to emphasise that primary (and, for that matter, secondary) elements are only taken up selectively by individual visitors, depending upon motives and the organisational characteristics of the visit.

Jansen-Verbeke's (1986) work is valuable in defining the range of urban facilities around which urban tourism tends to develop. Since in most towns and cities these facilities are normally spatially focused – often for historical and commercial reasons – the effect is to produce well-defined tourist spaces within which tourism itself becomes concentrated. These spaces are often coincident with historic cores and central business districts (Bull and Church, 1996). Within such zones, streets and sub-areas may develop associations with particular forms of provision – for example, the links between nightclubs and 'red-light' activity in London's Soho district or the Warmoestraat in Amsterdam; or the concentrations of London's theatres on Shaftsbury Avenue, its concert venues on the South Bank and its museums in South Kensington.

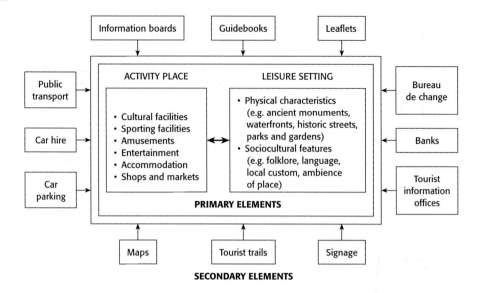

Figure 4.3 The structure of the urban tourist product
Source: Adapted from Jansen-Verbeke (1986).

The structuring of many cities into distinctive and well-defined tourist spaces also encourages concentrations of supporting facilities, especially accommodation. Except in situations where local planning controls limit associated tourism development (as is now evident in cities such as Amsterdam), hotels and other forms of tourist accommodation tend, for the convenience of the tourists, to concentrate in broadly the same areas as the key attractions. In London, for example, nearly 60 per cent of the hotels and almost three-quarters of tourist bed spaces are located in the three central London boroughs of Westminster, Kensington and Chelsea, and Camden. Elsewhere in the Greater London area, only Hillingdon and Croydon (which contain the key gateways of Heathrow and Gatwick airports) show significant concentrations of provision (Bull and Church, 1996).

The following case study illustrates the tendency towards concentration and the delineation of distinctive tourist spaces, by reference to a recent analysis of tourist districts in Paris.

Case study: Tourist districts in central Paris

Tourism to urban Paris is presently believed to involve around 20 million visits per year, with tourists attracted by the blend of historic sites, buildings and monuments; the world-class collections in the major museums such as the Louvre; a diverse range of entertainment, restaurants and bars; and a vibrant nightlife. Yet 'despite the importance of tourism in Paris . . . comparatively little attention has been directed to examining its spatial structure' (Pearce, 1998: 51). Pearce's (1998) study of tourist districts in Paris therefore represents an attempt to forge a better understanding of the functional linkages between tourists, facilities and space and to try to take the definition of tourist districts beyond the simple identification of patterns of supply.

The basic geography of tourism in Paris is shaped by the distribution of the major attractions (Figure 4.4), but by examining spatial patterns at a more local level – within areas that are generally recognised as containing significant levels of tourism – more subtle patterns and effects are revealed. Three contrasting areas were analysed:

1 The Île de la Cité (which contains the cathedral of Notre Dame).
2 Montmartre, a more peripheral and largely residential district, but with strong associations with Parisian café culture of the 1890s and a more recent reputation for nightclubs and sex shops.

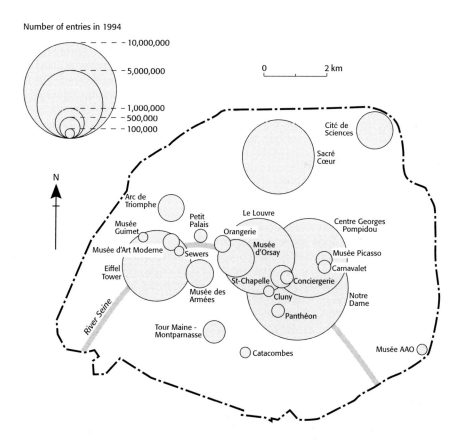

Figure 4.4 Visitor levels at major tourist attractions in central Paris
Source: After Pearce (1998).

3   The Opera Quarter, a zone of high-class retailing but also including the Louvre and the Tuileries Gardens.

Each district differs not just in the style of tourism for which it principally caters, but also in terms of the nature of the underlying functions. Part of the difference between tourist districts therefore stems from the nature of pre-existing development and its degree of adaptability in accommodating more recent tourist infrastructure.

The analysis shows that the functional definition of tourist districts may work at a number of geographical scales. For example, observation of the many guided tour parties visiting Notre Dame on pre-planned itineraries, suggests that for these groups the notion of a tourist district works only on a large scale and is loosely defined. The attention of these tourists tends to be directed only at the major sites (such as Notre Dame, the Eiffel Tower, or the Louvre) and that relatively little attention is focused on adjacent or intervening spaces.

In contrast, the analysis of tourist space at a district or local level points to some important ideas concerning how such spaces are defined and used. In each of the districts important attractions (such as the Sacré Cœur in Montmartre) act as magnets to draw tourists into the area, but spillover effects then encourage development of secondary attractions and facilities (such as shops, bars, restaurants and *bureaux de change*) adjacent to the primary attractions.

However, the development of the functional linkages between primary and secondary attractions is shown to be highly contingent upon access conditions and the degree to which intervening non-tourist places inhibit spontaneous exploration by obscuring or obstructing access to other nearby attractions. For example, on the Île de la Cité, significant visitor movements occur between the main coach park at the eastern tip of the island and the Place du Parvis which provides a primary access point to Notre Dame Cathedral. As a consequence, there is a concentrated development of secondary facilities along the Rue du Cloître Notre Dame and the Rue d'Arcole that take advantage of the passing trade (see Figure 4.5). However, the physical presence of the Préfecture de Police and the Hôtel Dieu (which is a hospital) discourages all but a fraction of tourists from visiting the Conciergerie and the Sainte-Chapelle – as Figure 4.4 reveals – even though the different attractions are in close proximity.

The importance of pedestrian corridors was emphasised in the other districts that Pearce investigated. In Montmartre, for example, linear agglomerations of sex shops, cafés and cabarets along the Boulevard de Clichy (which includes the Moulin Rouge) are a prominent feature in the micro-level patterning of tourism in the district, whilst in a different context, the clustering of duty-free shops aimed at tourists on the high-class shopping street of the Avenue de l'Opéra is also conspicuous. Interestingly, the pattern of tourist shops in the latter street shows a clear decline in the level of concentration as the distance from the primary attraction – the Louvre – increases. Field observation also showed that visitors moving off the main pedestrian routes and into adjacent streets would see a similar decline in the incidence and level of concentration of tourist provision and activity, emphasising the significance of access routes in shaping the microstructure of tourist districts.

Pearce's study does not, of course, address the behavioural patterns of tourists within the different districts beyond the point of basic observation or extrapolation from the pattern of supply, but it is valuable in highlighting

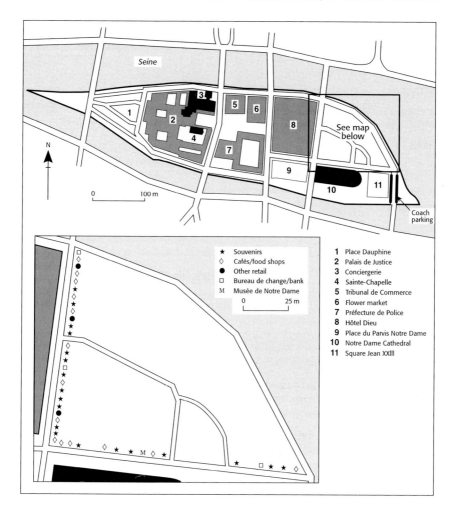

Figure 4.5 Tourist attractions and facilities on the Île de la Cité, Paris
Source: After Pearce (1998).

some of the ways in which agglomerations of tourist facilities are formed and identifies some important factors and processes that appear to regulate how space is used by tourists in a city and by which tourist districts may be defined.

Source: Based on information from Pearce (1998).

The observable tendency towards a spatial and organisational clustering of tourist functions within urban areas that Pearce identifies in Paris had earlier prompted Burtenshaw et al. (1991) to propose a simple descriptive model of the tourist city as a composite of more specialised tourist spaces (Figure 4.6).

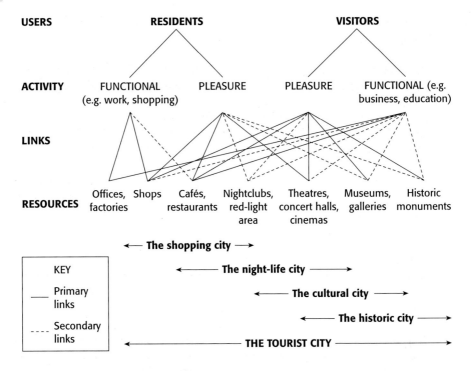

Figure 4.6 Functional areas and theoretical linkages in the tourist city
Source: Adapted from Burtenshaw *et al.* (1991).

This model illustrates how a range of local (i.e. resident) and non-local users (i.e. tourist) connect with different urban resources, and how these resources intersect and overlap to form principal components in the overall pattern. In the original model, the functional links are shown uniformly, but in reworking the idea (Figure 4.6), the links are differentiated according to an expected pattern of likely use by different user groups. The overlap between the various 'cities' has also been extended, reflecting the reality in which provision such as nightclubs and cafés form key elements in urban leisure cultures and even, perhaps, the manner in which shops form part of the city nightlife, either through extended opening hours or the enjoyment of window-shopping after hours.

This model has a value in providing a conceptual basis to understanding the organisation of tourism in city space, but in the context of the central argument in this book, there is an additional significance in emphasising the spatial coincidence of urban tourism and local recreation, at least within parts of the city.

Recreational space and activity

The complexity and diversity that characterise patterns of urban tourism apply with equal force to patterns of urban recreation and the resources on which recreational activities depend. This is a consequence of several factors, including:

- the relatively extended period over which recreational patterns have developed and the way in which successive 'layers' of provision have been established;
- the increasing diversity of demands and personal preferences that have emerged as personal lifestyles have become more leisured and the range of opportunities for urban recreation (both indoors and outdoors) more extensive;
- more recently, the development of urban tourism which has stimulated additional forms of provision and use that centre around tourism, but which – as already noted – draw in local recreational use as well.

It is impossible to locate with precision the point at which provision for urban recreation becomes an established aspect of urban development. Practices varied across space, through time and in the degree to which recreational space was placed in the public (as opposed to the private) domain. Medieval towns in Britain certainly contained spaces that were used for leisure (including street areas, market squares and common land) although, as Towner (1996) points out, their leisurely use was normally incidental to other (primary) functions. Equally, whilst the Renaissance in Europe promoted significant new ideas in urban design in which the more generously proportioned streets, squares and piazzas of the remodelled Italian cities were widely imported to fashionable parts of major cities such as London, Paris or Berlin, access to some of these spaces was often dependent upon social status. This was especially true when provision was set within exclusive residential developments, such as the Georgian squares of Bloomsbury in London, or the New Town in Edinburgh.

Organised, public provision for urban recreation tends, therefore, to be associated with the emergence of the industrial city. In the British context, Williams (1995) has suggested a simple, threefold phasing of development, beginning with a 'formative' phase from the mid nineteenth century and extending up to 1914. In this period the need for public provision of recreational facilities became generally recognised and by the end of the century a foundation of public provision (in the form of parks and public walks, recreation grounds, baths and libraries) had been laid down through a succession of Parliamentary Acts. This phase also saw the development of a distinctive commercial sector of provision, focused upon public houses and various forms of entertainment, such as music halls and popular theatre.

The second phase – 'consolidation' – Williams locates between 1918 and 1939. During this phase the physical expansion of urban areas under processes of suburbanisation initiated a parallel expansion of recreational space – for example within suburban parks – whilst the wider application of planning principles to residential development generally endowed homes with gardens. At the same time, concerns over public health and fitness encouraged a new emphasis upon sport – for example, through the introduction of public playing fields and the inclusion of sports pitches in suburban parks.

Finally, the period since 1945 is described as a phase of 'expansion'. Here, the growth in the overall demand for recreation was mirrored not only in an augmentation of existing facilities, but also in an extension into new areas of provision (such as indoor sports and leisure centres, large-scale entertainment complexes, shopping malls and urban heritage centres) that reflect the new leisure tastes in the post-industrial city.

This chronology of development helps to establish the notion of a progressive 'layering' of provision as recreational spaces in the city are built up through time (see also Burgers, 1995), but it is important to recognise that the process also translates into a distinctive spatial patterning of recreational facilities within urban areas (see, for example, Spink, 1989). This patterning reflects a range of basic factors, including not only the chronology (in which later phases of development tend to occupy sites that are further removed from the centre) but also:

- the space requirements of different types of facility;
- the ability of different recreational uses to compete within the urban land market;
- the accessibility requirements of different facilities;
- the typical behavioural patterns associated with usage – many of which are characteristically local.

Hence, for example, central areas often reveal significant concentrations of commercial recreation facilities that benefit from the accessibility associated with central positions and that can also afford to meet the higher costs of land. In contrast, recreational facilities that are more extensive in character and lack a commercial basis to their operations (for example, sports fields), typically locate towards the urban periphery. The main exceptions to this pattern are revealed in:

- recent trends towards the growth of out-of-town shopping malls, retail parks and entertainment complexes in peripheral locations;
- redevelopment of inner-city sites (such as docklands) as heritage attractions or zones of entertainment (see Chapter 6).

Table 4.2 sets out a model pattern (based upon observations of the organisation of recreational spaces in British cities) that might result. More importantly, perhaps, Table 4.2 also shows that in the contemporary urban place, recreational space is now an integral element within the urban form, and whereas tourist space occurs selectively, recreational space of one type or another is virtually ubiquitous.

Williams's (1995) approach emphasises physical forms of development and provision, but an understanding of how facilities are demanded and used requires a wider appreciation of the sociocultural context within which urban

Table 4.2 Distribution of recreational facilities and spaces within intra-urban zones – a model pattern

Zone or locality	Likely recreational spaces and facilities
Central area	• Small public gardens, sometimes larger parks • Public squares and street space • Retailing • Entertainment such as cinemas and nightclubs • Cultural facilities such as museums, galleries, concert halls, historic buildings • Restaurants and bars • Indoor leisure centres
Inner-city areas	• Parks (typically of nineteenth-century origin) • Recreation grounds and play areas • Street space • Some private domestic gardens • Allotments • Sports stadia, especially football grounds of professional clubs • Public houses • Community centres • Working clubs and institutes • Regenerated industrial/transport sites (for example dockland areas) that may include retailing, entertainment or heritage attractions
Suburban areas	• Suburban parks (typically of twentieth-century inter-war and post-1945 origin) • Recreation grounds and play areas • Playing fields • Sports pitches • Domestic gardens • Allotments • Street areas • Community centres • School facilities • Local service centres (which may include public houses and branch libraries, as well as indoor/outdoor sports centres in larger cities)
Urban fringe	• Domestic gardens and street areas • Out-of-town retailing, entertainment and cinema developments • Sports grounds • Golf courses • Sports stadia (through relocation from inner-city sites) • Quasi-rural open space (including urban woodland or commons) • Footpaths and bridleways

Source: Adapted from Williams (1995).

recreation (and tourism) have developed. Here the work of Clarke and Crichter (1985) is instructive in setting out how social, economic and political processes have influenced recreational patterns. Their analysis (which again focuses upon the British experience) identifies several relevant themes, including how:

- For working people, the transition in the early part of the nineteenth century from a pre-industrial to an industrial pattern of living created new levels of distinction between work and non-work.

- The increasing regulation of popular leisure during the early Victorian period produced a significant reduction in the opportunities to pursue traditional recreations but how, also, legislation (under some influence from the Rational Recreation Movement – see Chapter 2) began to create new opportunities through provision of public open space and civic amenities such as public baths, museums and libraries. This reflects a developing role for local government as a provider that becomes more significant with the passage of time.

- In the latter part of the nineteenth century, social differentiation brought on by the rise of the Victorian middle class created socially distinctive practices and precipitated new directions in working-class recreations centred around, for example, the public house; a rapidly emerging commercial entertainment sector; and (for men) the spectacle of professional team sport.

- By the early decades of the twentieth century, mass markets and associated institutional forms of provision had become characteristic, but so too had an emerging focus upon home-based leisure as living standards improved.

- After 1945 and, especially after 1960, urban leisure becomes even more family- and home-centred, as levels of domestic consumption rise whilst the role of public forms of leisure tends to decline. Within this period, too, whilst differentiation of leisure patterns according to class remains detectable, new patterns that cross-cut conventional social divides emerge – for example, in the development of urban youth leisure cultures and those of ethnic and minority groups. The post-1945 period also witnesses the maturation of a pattern of both governmental intervention (at national and especially local levels) and a commercialisation of leisure that has been progressing since the middle of the nineteenth century (Clarke and Crichter, 1985).

Urban recreation patterns in countries such as Britain draw, therefore, upon an extended pattern of physical development and associated sociocultural changes which have created modern patterns that are characteristically extensive and diverse. Some of that diversity is reflected in Table 4.3 which lists the major recreations and sports practised by people in Britain – as enumerated in the General Household Survey (GHS) (ONS, 1998) and supporting surveys of social trends (ONS, 2001). The picture that the GHS paints is a partial one, both in the selective nature of its coverage (with its emphasis upon sports or active recreations and in-home leisure), as well as its disregard for areas such as recreational eating and drinking, the use of entertainment and recreational shopping. However, other data can help to fill some of these gaps. (Although the data reflect the findings of national surveys, these are broadly indicative of

Table 4.3 Principal recreational interests and activities in the UK, 1997–9

Activity	% participating*
Watching television	99.0
Visiting/entertaining friends, etc.	96.0
Listening to radio	88.0
Listening to tapes/CDs, etc.	77.0
Walking	68.2
Reading	65.0
Cinema	56.0
Gardening	49.0
DIY	43.0
Swimming (indoors)	35.1
Theatre	23.0
Cycling	21.4
Keep fit/yoga/aerobics	20.7
Dressmaking/needlework, etc.	19.0
Cue sports (snooker/pool)	19.2
Tenpin bowling	15.5
Classical music concerts	12.0
Golf	11.0
Weight training	9.8
Darts	8.6
Soccer	8.5
Running/jogging	8.0
Tennis	7.1
Badminton	7.0
Ballet/opera	6.0
Fishing	5.3
Table tennis	5.3
Bowls	4.6
Squash	4.1
Cricket	3.3
Basketball	2.0
Rugby	1.3
Athletics	1.2
Hockey	1.1

* Participation period varies. GHS data refers to participation in the previous year, data derived from *Social Trends* (ONS, 2001) refer to participation 'these days'.
Sources: ONS (1998, 2001).

urban patterns given the dominance of urban populations within the sampling frames for the surveys.)

Several features of urban recreation are revealed in Table 4.3. First, the importance of the home as a recreational environment is reinforced through the extremely high levels of participation (and, in all likelihood, the frequency of participation too) in television viewing, entertaining and listening to radio and recorded music. Other home-based recreations such as reading, gardening and DIY are also prominent (see Glyptis *et al.*, 1987).

Second, the popularity of informal, active recreations – such as walking, cycling and swimming – is evident, although the data on cycling are inflated by the inclusion of some non-recreational use. Moreover, these recreations attract quite high frequencies of participation, with swimming recording an average frequency of participation for adults of over 8 times per year, and cycling over 11 times per year (ONS, 1998).

Third, commercial forms of recreation are widely enjoyed, especially cinema and, to a lesser degree, theatre and concerts. There are close linkages, too, between the commercial sector and the most popular sports, especially snooker and pool, tenpin bowling and darts. Not shown in Table 4.3, but very significant as commercial forms of leisure and recreation, is the incidence of recreational eating and drinking. Studies of the use of public houses and bars in individual cities, for example, suggest levels of participation that involve as many as 60 per cent of adult populations (e.g. Williams and Jackson, 1987), whilst the UK Day Visits Survey suggests that the main purpose of nearly 20 per cent of leisure visits to urban places is to eat or drink out (CRN, 1996).

Lastly, the data reveal the comparatively modest levels of participation in organised sports and physical recreations where, with the exception of golf, participation is generally in single percentage figures. However, given the overall scale of involvement in urban recreation, even single-figure percentages may translate into actual numbers of participants that exceed 1 million people.

Data of this type are valuable in providing a picture of recreational patterns at a fixed moment in time, but unless set with a time series of comparative figures, trends in participation cannot be discerned. When such comparisons are made, however, analysis suggests that some significant shifts in urban recreational tastes and preferences are occurring.

Changing patterns in urban recreation

The changing patterns in urban recreation (and the associated use of facilities) reflect several processes. There are, for example, some important shifts in the balance of usage between public and private space. These are evident in, amongst other changes: the widening use of the home as a recreational site; in an emerging trend towards the placement of children's play areas into commercial play barns (McKendrick *et al.*, 2000); in the dominance of the private sector in some expanding areas of sporting provision (such as golf); and, less obviously, the relocation of activities such as recreational shopping from publicly owned streets to privately owned indoor malls. Privatisation is also reflected in the pattern of regulation and control of recreational sites, particularly in the commercial/nightlife sector, and in the specific context of Britain,

Bennington and White (1988) have emphasised how the influence of New Right political agendas between 1979 and 1995 imposed a philosophy of privatisation on the delivery of many basic public leisure services.

These are important processes but additionally, two further areas of change merit emphasis:

1 The increasing significance of indoor recreational environments and activities.
2 A parallel decline in the relative attraction of some traditional outdoor forms of activity and provision.

The significance of indoor recreation in urban places

The development of indoor recreation is revealed in at least three distinct directions: in the growing significance of the home as a leisure environment; in the increasing attraction of some forms of commercial recreation (including leisure shopping); and in the development of a major sector of activity in indoor sport.

Home-based recreation

The rising significance of the home as a recreational space has been evident for some time (see, for example, Cherry, 1984; Clarke and Crichter, 1985; Roberts, 1989), and it is a simple truism that, collectively, more leisure time is spent at home than in any other category of urban recreational space. This is due to several advantages that the home environment offers, in particular: comfort, convenience, security and privacy. But it is also a consequence of cumulative affluence that characterised urban society in Britain, Europe, North America, Japan and Australasia at the end of the twentieth century and which has enabled much higher levels of equipment of the home as a place of leisure. Data for the UK for 1999 show that ownership of in-home electrical leisure goods has become almost universal, with 99 per cent of homes possessing a television, 86 per cent a video player, 72 per cent a compact disc player and 38 per cent a home computer. Sales of video tapes rose from 6 million in 1986 to 96 million in 1999, whilst sales of recorded music increased from 150 million units in 1973 to over 280 million units in 1999, with a marked shift from purchases of cheaper singles to more expensive long players and compact discs (ONS, 2001). Additionally, growing public interest in what may be termed 'lifestyle' recreations is reflected in rising levels of participation in domestic gardening and DIY (ONS, 1998), and home-based hobby interests have become more prominent in the recreational patterns as well.

The growth of home-based recreation is not, though, just a consequence of wider investment in home entertainment and leisure goods, but is also a product of progressive improvements in the home environment. For example, many homes – especially within the more affluent sectors of the urban community – have been extended spatially (often with conservatories as leisure rooms) and even in less prosperous areas, most homes now enjoy the benefits of central heating. This enables more effective recreational use of domestic space that historically was neglected – such as bedrooms or spare rooms – and permits the development of individual interests in ways that were often not possible when families were obliged to spend their free time sharing what may have been the one heated room in the house. James (2001) provides an interesting study of the role of bedrooms as leisure spaces for teenage girls, that illustrates the value of personal space within the home.

However, whilst these changes reflect broadly positive influences, Roberts (1989) also reminds us that for some people (for example, the elderly or the unemployed) the use of the home for recreation is often a consequence of constraints rather than choice – the trap of poverty, or of unemployment, or of fears of the wider urban environment.

Commercial attractions

The trend towards higher levels of use of indoor recreational space is also evident in the commercial sector. As noted above, urban places form important concentrations of commercial recreational facilities, and attractions such as pubs and bars, restaurants, nightclubs, concert halls and cinemas draw from a wide population base. Moreover, some of these sectors have seen very significant expansion in recent years. Cinema attendance, in particular, has benefited from the spread of multi-screen venues and a much-needed programme of modernisation and enhancement of facilities that had previously suffered from a cumulative cycle of declining audiences and outmoded infrastructure. In the UK, the proportion of the adult population that profess to visiting cinemas has risen from 34 per cent in 1987 to 56 per cent in 2000 (ONS, 2001), whilst in the USA, the number of cinema visits rose from 1.1 billion in 1985 to nearly 1.4 billion in 1997 (US Dept. of Commerce, 1999). A small number of American cinemas are located in the new generation of large-scale retail and entertainment malls (such as the Mall of America at Bloomington, Minnesota – see below), signalling quite clearly a new synergy between entertainment and indoor leisure shopping that is becoming a pervasive influence in American, British and European cities. (This theme is explored more fully in the final section of this chapter.)

Indoor sport

The third example of the rising significance of indoor recreational space is the movement indoors of a range of sports and physical recreations. This change has been largely stimulated by the growth in provision of multi-purpose indoor sports and leisure centres, but has also fed off rising public demand as a response to better awareness of issues of health-related fitness and the benefits of exercise. In England and Wales, provision of indoor sports centres rose from a low base of around 30 centres in 1972 to nearly 1,500 in 1995, whilst by 1998, public indoor swimming pools numbered over 1,300 (Sports Council, 1982; Sport England, 2002). Several of the popular recreations identified in Table 4.3 (including swimming, keep fit/aerobics, weight training and badminton) have all benefited from the increased level of provision. According to recent figures, more than 60 per cent of sport in the UK is now practised indoors, much of it within multi-purpose centres (ONS, 1998), although, of course, spectating (which is discussed below) remains an area of significant public participation out of doors.

As Williams (1995) explains, the modern indoor sports centre provides a number of appealing advantages over traditional outdoor provision. The environment is controlled, so participants are protected from the vagaries of weather and can play on surfaces that are normally clean, reliable and consistent. Second, by concentrating facilities, a range of activities can be accommodated in a single visit. Third, the social context of participation is enhanced by the provision of spectating areas for non-participants and spaces such as bars and restaurants that allow an extension of socialising around the sporting event.

The decline in traditional forms of outdoor provision

Set against this growth in indoor forms of recreation, there has been a decline in usage of some of the more traditional sites of outdoor recreation, such as parks, recreation grounds, playing fields and residential streets. The decline of urban parks has been especially representative of changes in public preferences and behaviours.

For much of the last 200 years, urban parks have provided an essential component in the public provision for outdoor urban recreation and, in major cities such as London, Paris or New York, form attractive and popular additions to the range of tourist spaces too (Figure 4.7). From early, and often tentative, origins in emerging industrial cities, parks of one form or another have become integral components in urban land-use patterns (see Chadwick, 1966 and Conway, 1991). They provide key sites for both active and, especially, passive forms of recreation for a very wide cross-section of the population, and their value as publicly provided, localised recreational spaces has been reaffirmed in a succession of studies of urban park usage (e.g. Bowler and Strachan 1976a, b; Walker and Duffield, 1983; Williams and Jackson, 1985; Burgess *et al.*, 1988; Welch, 1991; Page *et al.*, 1994; Greenhalgh and Worpole, 1995). The latter authors summarise the attributes and attraction of parks succinctly in stating that '. . . [parks] are local facilities; people who use them, use them frequently; they mostly walk to them; and they are accessible to all ages, and all walks of life' (Greenhalgh and Worpole, 1995: 3). (These qualities are

110 TOURISM AND RECREATION IN URBAN PLACES

Figure 4.7 The traditional image of the urban park – ornamental horticulture, seating and parkland sports (Stafford)

Table 4.4 Typical attributes of visitors to 10 urban parks in the UK

- Males outnumbered females at a ratio of 6 : 4
- Young adults and people in middle age made up more than 50% of all park users
- Around one-third of users visited alone
- Over two-thirds of users lived locally, taking less than 5 minutes to reach the park
- Almost 70% of users walked to the park
- Most users visited frequently and up to 40% said they visited on a daily basis
- Most visits are of a short duration, typically less than 30 minutes, although fine weather does encourage longer stays
- Common reasons for visiting included walking, exercising dogs, taking short cuts and taking children to play

Source: Adapted from Greenhalgh and Worpole (1995).

illustrated further in Table 4.4 which summarises typical patterns of use revealed in a study of 10 urban parks in Britain conducted by Greenhalgh and Worpole, 1995.) Furthermore, because parks often constitute local meeting places and focal points within urban communities, they may acquire a cultural significance in helping to define a local sense of place and an attachment to place (Crouch, 1995).

However, significant though parks might be, there is now ample evidence that points to a spiralling process of rejection of these spaces by a majority of urban residents, and whilst parks retain a value and significance for a residual group of users, these are now often in a minority. These changes owe something to the emergence of alternative (and competing) recreational spaces – including new patterns of natural green space in cities (see Nicholson-Lord, 1987, 1995), as well as the domestic or commercial leisure sites discussed above. But changing patterns also reveal a public rejection of a traditional facility that is often beset with problems in both management and use, and is widely perceived as anachronistic when set in the context of public tastes and preferences in post-industrial urban leisure.

The problems of urban parks are shaped by a familiar set of issues that Greenhalgh and Worpole (1995) summarise as follows:

- A lack of effective management that results from a combination of financial constraints on urban leisure budgets; the relegation of issues of public open spaces to the bottom of local political agendas; and a failure to address problems of siting and accessibility that are typical of older facilities set within an evolving urban landscape (see also Welch, 1991).
- Poor levels of maintenance, usually as a direct consequence of the problems of management, and a tendency to create a style of landscape based around areas of mown grass that possess limited amenity and aesthetic interest.
- Rising public concerns over the safety of public open space (see also Burgess, 1995), which links directly to the way in which parks often reflect the wider social pathology of cities, providing havens for vagrants and rough sleepers, drunks, drug users and people with mental disabilities.
- Problems of vandalism.
- Problems with dog fouling which Greenhalgh and Worpole (1995: 14) describe as an issue that, for many people, symbolises the decline of urban parks.

Of course, not all urban parks are in decline and there are many examples of facilities that have been effectively reinvented in new and exciting forms and which sustain high levels of use, often through programmes of planned and promoted events (see, for example, Williams, 1995: 173–7 on Telford town park in Shropshire, England – Figure 4.8). Similarly, in many cities other forms of public open space, especially street space in central areas, have enjoyed a renaissance as leisure sites through pedestrianisation and associated enhancement of the townscape. At the same time, though, ordinary residential streets and roads suffer increasing limitations on their traditional recreational roles as traffic levels rise inexorably (Hass-Klau, 1990). The underlying message is, therefore, clear. Many types of urban recreational space that possess a traditional significance are exerting a more selective and limited appeal within recreational patterns at the beginning of the twenty-first century and in their place, new patterns of consumption are emerging.

Figure 4.8 Part of Telford town park (Shropshire). Reclaimed from derelict land, the park is now actively promoted as a regional attraction within the English West Midlands

New urban places of recreation and tourism

The context in which the new patterns of production and consumption of urban leisure spaces are set is, of course, the emergence of the post-industrial city. The essential nature and implications of this process have already been set out in Chapter 1, but to try to illustrate how post-industrial shifts have produced new opportunities for convergence in urban recreation and tourism, two examples, the growth of leisure shopping and the role of sport as an urban spectacle, are outlined.

Leisure shopping

Newby (1992) asserts, quite correctly, that within urban populations there exist a range of shopping behaviours. These may be visualised as arranged along a continuum. At one pole there is the functional and routine acquisition of day-to-day necessities, at the other, the purely leisurely exercise of shopping as a recreation, although in many situations, of course, individual shopping trips may combine a range of behaviours from different areas of the continuum. (For a detailed insight into the routines associated with differing styles of shopping in two industrial cities – Manchester and Sheffield – see Taylor *et al.*, 1996.)

Not only does there exist a range of shopping behaviours, but also there is a diversity of shopping environments within which the opportunities for leisure shopping will vary. This idea has been conceptualised by Jackson (1991) in the diagram shown in Figure 4.9. This proposes a tripartite relationship between shopping and leisure:

- shopping *for* leisure goods, in which the purpose of the shopping trip is essentially functional, although the object is to buy goods that will be used for leisure purposes elsewhere;
- shopping *and* leisure, in which the trip combines visits to shops (again for largely functional reasons) with patronage of leisure facilities in the same geographical area – for example, a cinema visit;
- shopping *as* a leisure activity, in which the purpose of the visit is to spend leisure time in the act of shopping – browsing, window-shopping and deriving pleasure from the ambience of the shopping area and its activity.

In its latter form, shopping becomes a modern form of *flânerie* (in which participants stroll idly through shopping centres in order to be seen, to see others and to fill their time) and although the notion of shopping as leisure has gained only gradual recognition, it is arguably a particularly important and influential form of behaviour.

The model also shows that each of these forms of leisure shopping may be linked with likely shopping environments (here expressed as variation in 'retail

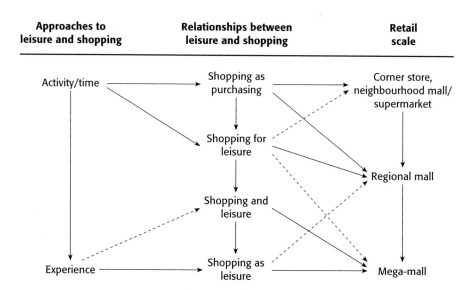

Figure 4.9 Theoretical relationships between shopping, leisure and place
Source: Jackson (1991).

scale'). Unsurprisingly, the mega-malls (such as the West Edmonton Mall, Canada; Mall of America, Minneapolis; Meadowhall, Sheffield (England) or Lotte World, Seoul) emerge as the context in which recreational motives will be most evident amongst shoppers. Tourism, too, is a conspicuous feature of usage of large-scale malls, although it is important to note that tourist shopping occurs in a range of other contexts as well, including: street markets; tourist shopping streets in central areas (as revealed in the case study of Paris earlier in this chapter); and in waterfront and previous zones of industry that have been redeveloped as places of amenity and tourist retailing. Ghiradelli Square (a former chocolate factory) and Pier 39 (a wharf) – both in San Francisco – are good examples of this latter pattern. Figure 4.9 also makes the important point that shopping in this latter situation is much more likely to be pursued for the intrinsic values associated with the experience, rather than as a simple functional activity.

Although the growth of leisure shopping – especially in malls – is widely presented as emblematic of the post-industrial transformation of urban life at the end of the twentieth century, it is interesting to note that the concept of shopping as a recreational use of time is not actually new. Nava (1997: 71), for example, reproduces advertisements for the fashionable London department store of Selfridges that date from 1909 and which make explicit reference to the opportunity that the store provided for shopping as 'a pleasure, a pastime [and] a recreation'. Furthermore, shops such as Selfridges rapidly acquired a reputation as one of the great sights of London which visitors to the city would expect to see. Visiting the store became 'an excursion' (Nava 1997: 69) and thus a part of urban tourism as well as local recreation. Nava (1997) also shows how as other forms of urban recreation became popularised in the 1920s and 1930s – especially cinema – recreational visits that combined shopping with cinema attendance became extremely popular, especially amongst women who often dominated the daytime cinema audiences.

Of course, such participation – particularly patronage of high-class department stores – was probably mediated by social class, and such recreational forms of shopping were often part of lifestyles where leisure was an attribute of social status. The key changes that have occurred towards the end of the twentieth century are that:

- shopping – as a recreation and as a key component of urban tourism – has become deeply entrenched in lifestyle across the social spectrum;
- it has taken on a new economic significance as leisure markets have grown;
- the retail landscape has become an important component in urban place promotion.

According to Goss (1993), in North America shopping is now the second most important leisure activity after television viewing, whilst for tourists, shopping is one of their most popular uses of time (Shields, 1992a) and stimulates extensive patterns of expenditure. A recent study of tourist shopping in Hong Kong, for example, suggested that as much as half of tourist expenditures in the destination went on shopping (Heung and Qu, 1998). Consequently, Jansen-Verbeke (1991) observes that shopping areas are now being widely developed as core elements in the urban tourist product because of the significant economic potential and the capacity to stimulate redevelopment of declining retail zones (Jansen, 1989).

The positional significance of (recreational) shopping in contemporary urban lifestyles owes comparatively little to the intrinsic appeal of the act of shopping as a traditional process of commodity exchange, as this is often a minor aspect of the visit to shopping areas (Shields, 1992a). Rather, its significance reflects the fact that shopping has become influential in defining key cultural attributes (such as identity) and in reflecting cultural values and associated ways of living. Clammer (1992: 195) writes that 'shopping is not merely the acquisition of things; it is the buying of identity . . . and the consumption even of "necessities" in situations where there is some choice, reflects decisions about self, taste, images of the body and social distinctions'. More widely, shopping (and the places in which it occurs) represents key elements in the emergence of consumer cultures in which routine acts (such as shopping) have become adjuncts of a wider quest for entertainment and amusement. Shields (1992b: 6) suggests that the significance of the shopping mall (as the epitome of new shopping practice and places) stems from its ability to offer 'a new spatial form which is a synthesis of leisure and consumption activities previously held apart by being located in different sites, performed at different times or accomplished by different people'. This view is echoed by Langman (1992: 40) who describes the shopping mall as 'the signifying and celebrating edifice of consumer culture' and by Hopkins (1991) who argues that in the North American context, at least, the mall is the principal forum for consumption and a popular place to spend money as well as (leisure) time.

Some of the inherent tensions and contradictions of contemporary consumer culture are reflected in the built design of the new generation of malls and urban shopping areas. One the one hand, sites such as the West Edmonton Mall draw on an eclectic blend of international styles, designs and traditions to create an overtly postmodern collage of references to other *places*, whilst simultaneously drawing upon a nostalgia that appears endemic in post-industrial urban communities for other *times* – both the industrial, and even the pre-industrial, age. Miller *et al.* (1998: 11) argue that shopping has become a part of the expansion of popular memory that underpins the burgeoning public interest in (leisure) heritage and draw attention to ways in which urban redevelopment has aided the expansion of sites of memory, especially within 'the carefully manicured centres of historic towns . . . [with their] . . . period shopping streets and precincts full of shops with small-paned windows and carefully calligraphed hanging signs' (see Figure 4.10).

This type of cultural repositioning of shopping helps us to understand the links between shopping, recreation and especially tourism. Shields (1992a: 102) describes contemporary shopping in malls as 'a social practice of exploration

Figure 4.10 The modern shopping mall is often consciously designed as an environment to attract tourists and leisure shoppers, as this example from Torquay reveals

and sightseeing akin to tourism'. The capacity for sightseeing and exploration in the mall is enhanced by sophisticated design techniques of illusion and allusion (Goss, 1993) that promote the spatial context in which shopping takes place above the simple status of 'backdrop', to an integral position within the construction of the overall shopping experience itself (Miller *et al.*, 1998). The inclusion of leisure facilities (such as cinemas, play areas or zones of entertainment); the widespread provision of facilities that encourage shoppers to linger (such as restaurants and food halls); the availability of guide and information services; in some mega-malls, the provision of hotels; the widespread use of water features and plants to create a relaxing mood – all conspire to provide a setting that is contrived for recreational and tourist use.

Lehtonen and Maenpää (1997) draw a number of parallels between mall shopping and tourism that reinforce this point. Both shopping and tourism, they argue, now take place outside the everyday spheres of home and work; both have an aesthetic relationship with their environment; and the core of their pleasure lies in the nature of the encounter with that environment. In addition, many of the malls that are designed around collages of images and spatial references to foreign and exotic places, succeed in the illusion because the shopping public itself possesses a dramatically expanded stock of place images to which they may relate – derived, in some measure, through tourism. In their most developed form, shopping centres become liminal places, i.e. sites that represent points of transition between normal social stations in which established norms are suspended or altered. They succeed partly because in their design they draw spatial referents to other liminal sites (such as seaports, resorts and tropical tourist destinations) around which they create a sense of the carnivalesque through their unique combinations of consumption and leisure (Shields, 1990; Goss, 1993).

Whilst the capacity of shopping centres to attract local recreational use has not been challenged, the scope for malls to shape tourism patterns has been less clear. Writing in 1991 about the West Edmonton Mall in Canada, Butler (1991) argued that malls were unlikely to shape tourist patterns, although they would – he conceded – attract increasing numbers of visitors who were already in the area. Subsequently, however, a growing body of information has suggested that the largest malls do, in fact, possess a capacity to act as destinations for tourists and to shape patterns of travel. A recent study by Craven (2000) of the Bluewater shopping centre in north Kent (England) showed that amongst the 25 million visitors that are currently estimated to visit the centre, were noticeable numbers of French and Belgian tourists taking advantage of enhanced cross-Channel ferry and rail services (see Chapter 3). In the largest malls, the tourist presence is even more substantial. The following case study of the Mall of America reveals the potential scale of tourist visiting to major shopping malls and aims to illustrate, particularly, the integral nature of leisure facilities and opportunities in sites of this nature.

Case study: Shopping malls as sites of recreation and tourism – the Mall of America

Appropriately, perhaps, the site of the Mall of America at Bloomington, Minneapolis, was itself formerly a leisure site. Until they relocated to an indoor arena in the downtown district of Minneapolis in 1982, Minnesota's professional baseball and American football teams (the Twins and the Vikings) played on the site at Met Stadium. The 31-ha site enjoyed excellent access to major highways and the international airport and whilst a range of possible alternative uses were considered (including office space and residential land), a retail and entertainment centre was eventually selected as the preferred form of redevelopment. Work commenced in June 1989 with the complex opening three years later, in August 1992.

The design of the lower level of the Mall is shown in Figure 4.11. The Mall is 'anchored' by four nationally recognised department stores: Bloomingdale's, Macy's, Nordstrom and Sears and additionally there are over 500 smaller retail units in the complex, arranged over three levels. Each of the retail 'avenues' has

The Sport Tourism International Council (summarised in Hinch and Higham, 2001) have identified five categories of sport tourism:

1 Attending or participating in major sporting events.
2 Visiting attractions (such as heritage sports facilities, sports museums, or major sporting venues).
3 Resorts with a sporting focus.
4 Cruises that centre around sporting themes or personalities.
5 Sports tours or activity holidays.

In the urban context the most significant potential for sporting tourism is within the first two of these categories: attending sporting events and visiting sites and attractions associated with sport.

Event-based sports tourism operates at a variety of scales – from mega-events (such as the World (Football) Cup or the Olympic Games), to the more routine support of professional sports teams. Large-scale events have begun to attract increasing levels of interest in urban policy-making, primarily because of the presumed benefits in terms of urban regeneration, new facility provision and image promotion that hosting such events is believed to bring (Hall, 1997). Consequently, competition to host major events such as the Olympic Games has become intense in light of the supposed advantages (Hiller, 2000), and even unsuccessful bids often result in an enhanced level of recognition for cities in question (Whitelegg, 2000). However, where an event fails to perform as an attraction at the expected level – as occurred at the World Student Games in Sheffield in 1991 – a rather different legacy of urban debt is often created. New facilities provided for the major event may enhance local recreational opportunity, but some authors have questioned the long-term value of such provision. The scale and associated running costs of prestige sports facilities are often significantly out of step with local recreational needs (Whitson and Macintosh, 1996).

Large-scale events are designed to bring significant numbers of visitors (who constitute tourists) to the city in question. The 1992 Olympic Games at Barcelona, for example, were expected to draw in excess of 500,000 visitors to the city (Redmond, 1991). However, in the case of more routine forms of spectating (such as professional team sport), the pattern is rather different. Here, traditional patterns show that whilst major teams (such as Manchester United AFC) may draw support from extended catchments and a nationwide network of supporters' clubs (Bale, 1993), most support is generated locally. Supporters of visiting teams do, of course, constitute visitors from outside, but they seldom stay in the host city for more than a few hours. Their visit is, therefore, more recreational than touristic in character.

However, in some situations the pattern is different. In the USA, for example, the effects of increased distance between major urban centres often means that supporters travelling to follow professional football, basketball or baseball teams have to build their journeys into weekend breaks at the cities they are visiting. Of the 333 million overnight leisure trips monitored by the Travel Industry Association of America in 1997, some 6 per cent (nearly 20 million) were made to attend sporting events – mostly as spectators (Loverseed, 2001). Furthermore, as the organisation of some sports takes on an increasingly international dimension – for example, in the development of football's European Champions League – so the scope for supporters to spend several nights away from home in visiting foreign cities is increased. In so doing, the supporters become urban tourists.

Data on the scale of such activity are elusive and the numbers who travel abroad in support of teams are likely to be relatively low at present, but figures for attendance at regular domestic professional sporting fixtures point to the scale and significance of sport as an urban spectacle and hence, the potential for expansion in this sector. Stevens (1995) suggests that on a typical British Saturday in January, over 750,000 people will be inside urban sports stadia watching live sport. Annual attendance in the English football league (including the Premiership) in the 1999–2000 season approached 25 million, and whilst many of these people are repeat visitors who watch soccer regularly, weekly average attendance exceeded 600,000 (AFS, 2002). In the USA, the larger population base generates even greater aggregate levels. In 1997, for example, attendance at major league baseball reached nearly 65 million people, whilst the men's collegiate basketball league drew 21.6 million and the collegiate American football leagues attracted 36.8 million spectators (US Dept. of Commerce, 1999).

However, not only is the performance of sport a major recreational and tourist attraction, but also the sites of that performance and locations that celebrate urban sport are becoming visitor attractions. This is evident in both the placement of sporting stadia onto recreational and tourist visitors' itineraries, and the burgeoning industry in sporting museums and halls of fame.

The concept of the hall of fame as a device for celebrating the contribution of notable individuals to a walk of life originated in the USA in 1901, and has permeated American culture to the extent that more than 400 were in existence across North America by 1990 (Redmond, 1991). Of these, nearly a quarter celebrate sporting achievement and the most popular draw around 250,000 visitors annually (Stevens, 1995). These form part of a broadening base of heritage attractions that recognise the cultural significance of urban sport that also embraces a growing number of museums. These have developed in a number of contexts: sometimes they occur as celebrations of particular sports (such as the British Golf Museum at St Andrews in Scotland, the National Horse Racing Museum at Newmarket, or the Cricket Museum at Lord's, London); on other occasions sports museums become attached to sporting clubs. In Britain, Manchester United, Liverpool and Arsenal football clubs all have club museums at their home grounds.

Club museums are one factor that attract visitors to sports stadia, but it is evident that as the technology of sports stadia construction develops, so the buildings themselves take on a new significance as truly spectacular landmarks that draw the tourist gaze as attractions in their own right. In some cases tourists have been visiting sports stadia with strong historic or cultural significance for some time, albeit in relatively small numbers. The old Wembley Stadium in north London – which offered regular guided tours – was a case in point. More recently, however, stadium developments have begun to emerge that combine recreational, tourist and sporting interests in quite new ways. This has been especially evident in North America where a new generation of domed stadia have sought to mix provision for professional field sport with hotel accommodation, catering, conference facilities and retailing to create facilities that not only serve local sport, but are also highly attractive to recreational and tourist visitors (Bale, 1994). For example, the Toronto Skydome (which is a home to both baseball and football) accommodates 53,000 spectators and has 8 restaurants, a fitness club, a small movie theatre, several retail outlets and a 348-room hotel – including 70 rooms overlooking the synthetic 'field'. Daily guided tours aim to satisfy the curiosity of those who come simply as sightseers. As Bale (1994) indicates, such stadia share all the ambiguities of the shopping mall in their mixing of functions and facilities, whilst commodifying an essentially simple recreational tradition – the enjoyment of sport as a spectacle – into something much more complex.

Conclusion

Conventionally, towns and cities have been viewed as primary sources of tourists (rather than destinations) and locations for a broadly traditional range of local recreations, many of which were practised within clearly defined times and places that were familiar to a majority of citizens. The discussions within this chapter have tried to show that such notions can no longer be sustained. It is evident that cities (in particular) have become major national and international tourist destinations and their appeal is shaped (in part) through the increased diversity of recreational and tourist sites that urban places are now able to offer. This, in turn, reflects key processes of change in the position of leisure within the post-industrial urban lifestyle and the role of local recreation in many contemporary urban lives. These processes have seen the gradual eclipse of many of the conventional urban facilities around which urban leisure was formerly shaped (such as town parks) and in their stead, the emergence of new sites of consumption (such as shopping malls) in which leisure, recreation and tourism are being blended in novel and exciting ways. We cannot, therefore, draw simple conclusions about the role that urban places play in contemporary recreation and tourism or the times and places in which such activity occurs. Instead we must acknowledge the growing complexity of the interrelationships between urban recreation and tourism and the ways in which these key areas of leisure are forging new areas of shared spaces, practices and identities.

Questions

1. To what extent may the recent growth of urban tourism be characterised as simply the latest phase in an extended process of tourist use of urban places?
2. Compare and contrast the manner in which recreational and tourist space in typical contemporary cities is organised and arranged.
3. What have been the principal trends in the development of urban tourism since 1970?
4. Why has shopping developed so strongly as an area of recreational and tourist interest?
5. What evidence is there to support the notion that urban sport is now a distinctive component in urban tourism?

Further reading

There are several good, recent texts on urban tourism patterns and processes, including Law, C.M. (1993) *Urban Tourism: Attracting visitors to large cities*, London: Mansell, and (1996) *Tourism in Major Cities*, London: International Thomson Business. Page, S.J. (1995) *Urban Tourism*, London: Routledge is also well worth reading.

Texts on urban recreation are rarer, but a recent discussion is provided by Williams, S. (1995) *Outdoor Recreation and the Urban Environment*, London: International Thomson Press.

There are several texts on urban shopping, but two are recommended: Miller, D., Jackson, P., Thrift, N., Holbrook, B. and Rowlands, M. (1998) *Shopping, Place and Identity*, London: Routledge, and Shields, R. (ed.) (1992) *Lifestyle Shopping: The subject of consumption*, London: Routledge.

Sport is a neglected topic in discussions of tourism. The best recent works on the wider role of sport in urban communities (and which include brief discussions of tourism) are: Bale, J. (1993) *Sport, Space and the City*, London: Routledge and (1994) *Landscapes of Modern Sport*, Leicester: Leicester University Press.

Tourism and recreation in the countryside

CHAPTER 5

Introduction

The modern countryside, especially within westernised, industrial nations such as the UK, France or the USA, offers a deceptively complex context for recreation and tourism. Since 1945, recreation and tourism have become key components within a complicated and fluid set of changes that are affecting many rural areas, yet their influence has often been inconsistent and unpredictable. The context of change essentially centres upon tensions promoted by the diversity of demands that are now being placed upon rural spaces, within what has become 'contested' countryside (Cloke and Little, 1996). In part, these tensions reflect a reduction in the conventional productive functions of rural areas (especially within agriculture) and an increasing role for the countryside as an area of residence and amenity. Within member states of the European Union (EU), for example, presumptions in favour of agriculture have been eroded in the face of wider use of production controls and incentives to diversification of farm enterprise, including provision for leisure and tourism. This reflects a widespread recognition of the potential value of recreation – and especially tourism – in the economic realignment of rural areas.

But tensions have also arisen through the growing diversity of social constructions placed upon 'countryside', in which different groups and individuals attach differing values and expectations to rural areas. In particular, any understanding of the contemporary countryside must acknowledge the symbolic values that urban–industrial populations ascribe to rural places. Clark *et al.* (1994), writing specifically about England, argue that the countryside has taken on an important place in English self-understanding, a position reinforced through popular attachment to a deep-seated association between rurality and notions of tradition and stability. When such attachments are combined with the enhancement of the role of leisure in the construction of social identities that has occurred in post-industrial society (Kelly, 1983), the value and significance of the countryside as a recreational environment are clearly emphasised. Bunce (1994: 139), writing on Anglo-American perspectives, reinforces the point in observing that 'the use of the countryside for recreation can be seen both as an expression of broader cultural values and sentiments and as a fulfilment of basic physical and psychological needs'. However, whilst some constructions of the countryside explicitly promote its use for leisure, other perspectives – especially those that reflect greater public sensibilities to problems of the environment – are more likely to emphasise issues of conservation and control over leisurely use of rural land, than its active development.

The patterns of use of the countryside for recreation and tourism, although superficially evident, are also ambiguous. The visibility of recreational usage of rural areas – especially as revealed in the commonplace incidence of congestion of tourists and recreationalists at the popular sites – emphasises the appeal of the countryside, yet conceals the fact that the countryside is not a leisure environment for all. In Great Britain, for example, around 40 per cent of the population do not visit the countryside at all in a typical year (CRN, 1996). Amongst those who do use the countryside, the majority of visits are actually composed of high-frequency, short-duration, localised trips by comparatively small numbers of people who remain largely undetected in the site-based surveys on which so much of our knowledge of rural recreation patterns has been built. The Countryside Commission's National Survey of Countryside Recreation for 1990 (Countryside Commission, 1995) shows that 61 per cent of visits to the countryside were generated by just 11 per cent of the population. In practice, therefore, the use of the modern countryside for leisure is a complex blending of localised patterns of leisurely use (including rural residents as well as those of adjacent urban areas), with more irregular forms of recreational visiting and a growing number of rural tourists.

In this chapter we will reflect upon these key themes, looking especially at the ways in which the patterns of use of rural areas by both recreationalists and tourists have evolved from sporadic and often place-specific patterns of visiting, to the point at which recreation and tourism have become widely influential in shaping the form and character of rural areas and integral elements in many rural economies.

Evolution of recreation and tourism in the countryside to 1945

As in many forms of recreation and tourism, mass participation in countryside activities is largely a product of the post-1945 and, especially, the post-1960 period, but the origins lie much earlier. Pre-industrial rural communities across most of Europe enjoyed surprisingly rich patterns of locally based recreations and these were commonly supplemented by more occasional holiday events such as annual parish feasts, harvest festivals and wakes. These holidays provided valued opportunities for communal participation in traditional recreational pleasures such as music, dancing, races, wrestling and cock-fighting (Towner, 1996) and although subjected to periodic attack – such as during the growth of Puritanism in seventeenth-century Britain (Towner, 1996) or sabbatarianism in the early nineteenth century (Cunningham, 1980) – these traditional forms of recreation revealed a high level of resilience. However, the onset of the Industrial Revolution and the eventual emergence of urbanism as a dominant way of life, significantly altered the picture.

The seeds of change probably lay in the growth of capitalist economies in the sixteenth century which, in the countryside, led to the creation of new social hierarchies centred on country houses and their estates (Bunce, 1994). Not only did these estates become associated with new patterns of elite, leisure-based lifestyles in which sports such as shooting and hunting flourished but also, and more critically, with the wider appropriation of rural space to support exclusive forms of leisure. Curry (1994) notes that the enclosure movement in Britain removed large areas of common land from public access whilst the development of landscaped parks around country houses and the extension of legal controls over access and activities such as hunting, placed further limitations on popular recreation in the countryside.

Local recreations did, however, survive and it is interesting to note that from the early phases of industrialisation, local patterns were augmented by small but significant incidences of the use of rural areas for recreation by working populations in industrial towns. Curry (1994) explains such patterns as a simple and understandable quest for fresh air and open space as relief from the crowding and pollution of the new industrial towns – motives that apply with almost equal force today as then. For example, at Preston in northern England, the enclosure in 1833 of part of Preston Moor to create an early example of a municipal park (Conway, 1991) formalised a pattern of *de facto* local use by the expanding industrial population that had been evident for some time.

By the early decades of the nineteenth century, recreational patterns in some rural areas were also supplemented by early forms of tourism. Domestic touring was not an innovation – as the travel diaries in Britain of Celia Fiennes (1680s), Defoe (1720s) and Cobbett (1800s) reveal – but shifts in the tourist gaze during the latter parts of the eighteenth and the early nineteenth centuries onto picturesque and then romantic landscapes provided a major catalyst to the development of rural tourism. As Corbin (1995) has explained in the context of European seaside tourism, the eighteenth century was a critical period in the advancement of understanding of natural systems and relationships between people and nature, and this fostered new levels of interest and new ways of looking at scenery and landscapes, especially within science, art and literature. In particular, mountains and wilder regions that had previously been disregarded, took on a new-found appeal as affluent members of society began to visit such places in search of the sublime (Bunce, 1994). In Britain, aided by early examples of travel guides (such as William Gilpin's celebrated series of 'Observations' on picturesque regions such as the Wye Valley) and carrying the accoutrements deemed necessary for the proper appreciation of the picturesque (such as the Claude Glass – a device for viewing and framing scenes), a small but growing body of tourists visited areas such as Wales, the English Lake District and Scotland (Andrews, 1989).

Similar patterns of touring emerged in Continental Europe and in North America. In France, the writings of Rousseau, in particular, stimulated an emerging taste amongst Parisians for visiting accessible countryside (Green, 1990), whilst in the USA, wilderness areas such as the Catskill Mountains began to develop as a popular rural area for tourists. Initially, visits to the Catskills formed one element in the annual social season of fashionable New York and Boston society that also embraced spas and seaside resorts (Demars, 1990), but as travel on the Hudson River improved and the first hotels began to appear in the 1820s, so accessibility became wider in both a physical and a social sense (Johnson, 1990).

At first, such excursions tended to be focused upon a comparatively small number of favoured locations, and visiting such sites was exclusively the domain of those who could afford to travel and were sufficiently well educated to be both aware of, and responsive to, new tastes in art and literature. Wordsworth's often repeated vision of the Lake District as 'a sort of national property, in which every man has a right and an interest who has an eye to perceive and a heart to enjoy' was actually a highly exclusive vision – encouraging notions of access for people of taste but rejecting ideas that such countryside might be appreciated by artisans and the labouring classes. However, as the nineteenth century progressed, the new middle classes in Britain, Europe and the USA became conspicuous consumers of countryside recreation and after about 1870, the habit had broadened to include many skilled workers and tradespeople (see Figure 5.1). Poole (1983) described how in northern industrial towns in England, the industrial 'wakes' holidays commonly included excursions to accessible and attractive countryside (such as the Derbyshire Peak District or the Lake District) as well as the popular coastal resorts, especially Blackpool. In France, as early as the 1840s, areas such as the valley of Montmorency, north of Paris, saw regular invasions of working people from Paris at holiday times – mostly as day trippers, but sometimes as staying tourists (Green, 1990).

The scale of urban use of the countryside for recreation and tourism grew as levels of affluence, holiday entitlement and especially mobility improved towards the end of the nineteenth century and in the years immediately before the First World War. At this time, the bicycle was an important innovation in bringing wider access to the countryside to urban populations. The development of the chain-driven 'safety bicycle' and the pneumatic tyre prompted significant growth in cycling as a basis to both recreation and tourism in the countryside, particularly in the 1890s. In Britain, membership of the Cyclists' Touring Club rose from just 148 members in 1878 to over 60,400 in 1899 (Lowerson, 1995) and although the cost of machines at first limited participation

Figure 5.1 The picturesque landscapes that modern visitors to the countryside take for granted as areas of recreational or tourist attraction in practice only became fashionable through significant shifts in public taste towards the end of the eighteenth century (Loch Creran, Scotland)

mainly to younger professional men, demand stimulated reductions in the cost of bicycles that encouraged significant growth of interest in cycling amongst working populations by the outbreak of war in 1914. Similar (though not identical) changes were observed in the USA (Tobin, 1974) and on the continent of Europe, as the following case study of the early development of cycling in France reveals.

Case study: Cycling in France before 1920

In France, as in Britain and the USA, the appeal of cycling to the urban populations of the late nineteenth century lay in the new-found personal mobility that a bicycle provided, not just for routine journeys within towns and cities but especially as a means of escape to the countryside during periods of leisure. The fashion grew rapidly – estimates suggest that in 1884 there were perhaps 4,000 bicycles in use in France but within 10 years the numbers that were formally registered and taxed had risen to over 132,000. By 1913 the official figure had passed 3.5 million, although the true figure may well have exceeded 5 million with allowance for the many who did not register their machines to escape the tax.

The bicycle provided a means both for local visiting of country districts from cities such as Paris or Lyon, as well as extended forms of tourism into the remoter countryside which involved nights spent away from home. These wider excursions were aided by the improving state of roads and also by the extensions in the rail networks that allowed cyclists to transport their machines by train to gain access to areas that were previously remote, such as Brittany and the Massif Central. Touring by bicycle also benefited from active promotion by clubs such as the Touring Club de France (TCF) which not only provided guidebooks and information to members on routes and approved hotels, but also campaigned for improvements to roads and provision of roadside services. In 1913, more than 136,000 people were members of the TCF.

However, participation in cycling – whether as a local visitor to the accessible countryside or a more adventurous cyclo-tourist – was at first limited socially to younger members of the professional classes, people who had the energy to pedal the heavy machines and who could afford the time and the money to purchase and use a bicycle. In the 1880s, safety bicycles imported from England cost as much as 500 francs (when most industrial workers earned an average of only 3 francs a day), and although the combination of healthy demand coupled with improvements in production technology drove down the price to as little as 125 francs in 1905, cycling was still not easily afforded by ordinary people.

Initially, cycling was also predominantly a male pastime, despite the expectations of some that the freedom that bicycles provided would have an emancipatory influence on female leisure patterns. The evidence from France is only fragmentary but, for example, of the 1,138 members of the TCF in 1891, only 14 were women, and it seems likely that active participation in cycling in France by women remained relatively low until after 1918. By then, however, the boom in cycling as a fashionable form of recreation had passed, although its appeal as a cheap form of local travel to rural areas continued, especially amongst working-class citizens for whom the bicycle had at last become an affordable item.

Source: Based on information from Holt (1985).

The use of bicycles formed one element in a broader movement towards recreation and tourism in the countryside that developed strongly in the first few decades of the twentieth century and which also embraced activities such as walking, rambling and fishing. As with cycling, rambling was at first restricted to those with the time and the money to travel to rural areas (usually by train), and many of the first walking clubs (such as the Sunday Tramps in London or the Rucksack Club in Manchester) were solidly middle class and male in their composition (Walker, 1985). However, by the early years of the twentieth century – and especially in the northern industrial cities in England – working-class rambling clubs such as the Sheffield Clarion Ramblers (founded 1900) were attracting large memberships (Walker, 1985). Angling grew similarly in popularity and although the sport was socially differentiated – with game fishing for trout tending to be the preserve of the middle and upper classes – coarse fishing became popular with working men. The quest for waters to extension of earlier trends, albeit with a marked intensification. Within this process we may identify at least four key themes that we may use to structure an understanding of the changes that have occurred:

1 Growth in the levels and frequency of participation.
2 Diversification of activities and the spaces that are used for rural recreation and tourism.
3 Increasing commercialisation and commodification.
4 Progressive integration of recreation and tourism into the wider framework of rural production and consumption, especially in agriculture and forestry.

Growth in the level and frequency of participation

Contemporary patterns of recreation and tourism in the countryside comprise three distinct elements: day trips from home (which are mainly recreational); day trips from holiday bases; and staying visits in rural areas (the latter two both being primarily forms of tourism). As populations have become more affluent, leisured and, especially, mobile, the fact that each of these sectors has grown substantially is not in doubt. However, the true extent of the change is unknown, even though a growing array of sample figures and survey findings, both individually and collectively, confirm the contemporary significance of rural recreation and tourism.

In Britain, the earliest surveys of countryside recreation generally date from the 1960s, but their sporadic and often small-scale nature afforded only a sketchy picture of the levels and character of visiting. On the basis of this evidence, Coppock and Duffield (1975) concluded that perhaps half of the population made some use of the countryside for leisure. By the late 1980s, however, more structured and comprehensive surveys by the Countryside Commission (1985, 1995) were able to reveal with greater precision the levels of countryside use. These surveys suggested that perhaps as many as 70 per cent of the UK population now visited the countryside at least once in a year, and embedded within this aggregate figure was a growing number of frequent and regular users who generated a significant number of visits. Thus, recent UK Day Visits Surveys (CRN, 1996) have estimated almost 1.3 billion day visits to the British countryside, most of which were made from home, but with some 70 million made from holiday bases by tourists.

Visits within this latter category occur more selectively as a consequence of spatial variations in the incidence of domestic tourism, but in regions where important tourist destinations lie adjacent to attractive countryside, the overlap between local recreation and tourism may be pronounced. As an example, Figure 5.2 illustrates the composition of visitor flows to Dartmoor National Park from a range of contiguous source areas that include important zones of urban seaside resorts in the Torbay and Teignbridge districts. It reveals that in the majority of cases, flows of visitors comprise both local recreationalists and tourists staying in the region.

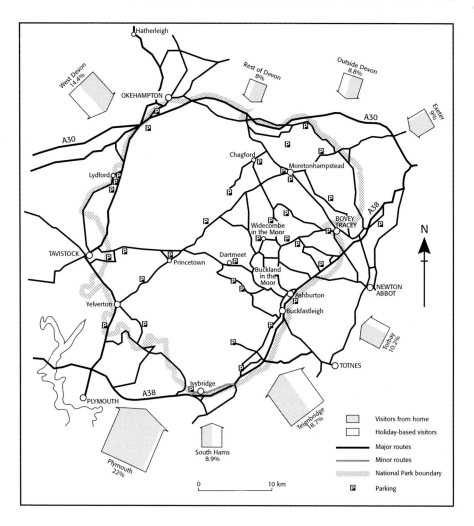

Figure 5.2 Origins of visitors to Dartmoor National Park
Source: Data provided by Dartmoor National Park Authority.

Evidence of the growth of staying holidays within the countryside is rather more elusive. The English Tourism Council (2000) estimate that the British generated some 65 million domestic trips involving at least one night away from home in 1998. However, the destinations of these trips are not disaggregated below the level of the tourism regions, so the proportion that are rural is unclear. Predominantly rural regions (such as Cumbria or Northumbria) tend to attract a relatively small percentage share of the overall market (4 and 3 per cent respectively), although even a small share represents a significant local trade. (On these data Cumbria attracts 2.6 million staying visitors annually and Northumbria 1.95 million.) Additionally, though, there will be substantial elements of rural tourism contained within visitor patterns in the major British tourism regions – especially the Heart of England (which attracts 8.5 million staying visitors) and the South West (which draws 15 million) – although the proportion of these visitors that select countryside destinations rather than urban places and seaside resorts is not clear.

One of the conspicuous recent developments in British domestic tourism has been the growth of short-break holidays (one to three nights), where the market grew by 23 per cent between 1989 and 1997 and generated some 37 million trips in 1997 (Beioley, 1999). Here the rural component has been more precisely enumerated, with 26 per cent of short breaks being to the countryside and a further 16 per cent focused on 'small towns', many of which will be semi-rural in character and location. This provides some indication of the relative attraction of rural destinations to staying visitors within the UK.

A similar picture of growth in participation is evident in the USA. The 1960 survey of the Outdoor Recreation Resources Commission (US Dept. of Agriculture, 1962) revealed that just over half the American population made at least one visit to the countryside in a typical year and, on that basis, predicted significant subsequent expansion that has duly materialised. For example, visits to national forests in the USA rose from 230 million visitor days in 1980 to over 340 million visitor days in 1996, whilst annual visits to the American national park system numbered over 275 million by 1997 (US Dept. of Commerce, 1999).

However, although the evidence points to a growing appeal of rural recreation and tourism, it also shows that appeal to be selective. Clark *et al.* (1994: 31) comment that 'a body of social survey evidence [in Britain] shows that the majority of regular visitors to the countryside are rural or suburban, white, home-owners, well-educated, and employed in non-manual occupations'. This distinctly middle-class profile of countryside visitors is quantified in the UK Day Visits Survey (CRN, 1996) which notes that countryside visitors revealed the highest proportions from the upper ABC1 social groups; the highest proportions in full-time employment; and, at 77 per cent, the highest levels of car ownership (Figure 5.3). Car ownership, in particular, emerges as a key to participation in contemporary rural recreation and tourism – generally an essential component in all but the most local of trips. This is a reflection of the deficiencies of public transport systems in rural areas of countries such as Britain, as well as the clear advantages of flexibility and convenience that the car bestows on its users.

Not only is the use of the countryside uneven in terms of the social composition of visitors, it is also variable in space and time. Kay (1996), in a comparative analysis of the Countryside Commission's National Surveys of Countryside Recreation and the UK Day Visits Surveys, shows not only the seasonal contrasts in visiting (Figure 5.4), but also – as revealed in the 1984 National Survey – the pronounced unevenness of use between weekdays and weekends (Figure 5.5). Unfortunately, comparison with more recent data is not possible as the UK Day Visits Surveys do not show patterns of visiting on

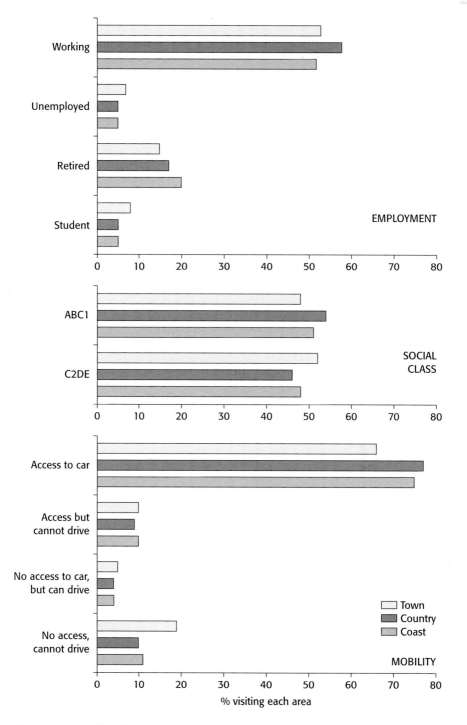

Figure 5.3 Profile of day visitors to urban, rural and coastal destinations in the UK
Source: Based on information from CRN (1996).

TOURISM AND RECREATION IN THE COUNTRYSIDE

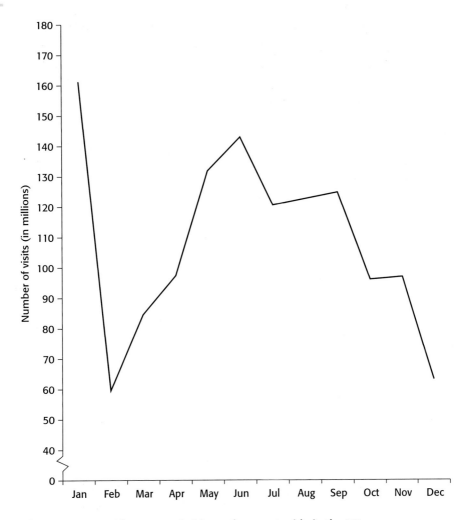

Figure 5.4 Monthly pattern of visits to the countryside in the UK
Source: Based on information from CRN (1996).

individual days. At the aggregate level, the 1994 survey showed that there were more countryside visits in the week than at the weekend, but when weighted on the basis of visits per day, the weekend proves to be more than twice as popular as the week as a time to visit the countryside (CRN, 1996). Thus, whilst the congestion associated with the relatively few days of peak usage attracts concerned attention, Kay concludes that 'the countryside's capacity to provide for enjoyment is grossly under-used for the greater part of the year' (1996: 22). The analysis also shows that much of the usage is local (see Figure 5.6) – although there is a significant element of longer-distance visiting that probably reflects the rather different patterns of rural tourism, as well as recreational day visits.

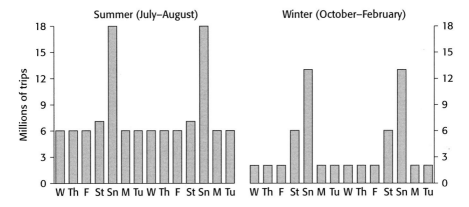

Figure 5.5 Contrasting temporal patterns of countryside visiting in the UK
Source: Based on information from Countryside Commission (1985).

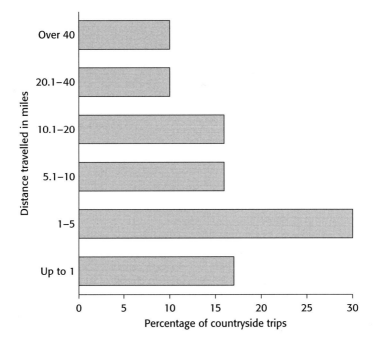

Figure 5.6 Distances travelled on countryside trips in the UK
Source: Based on information from CRN (1996).

The localised nature of countryside use is confirmed in the UK Day Visits Survey (CRN, 1996). This revealed that the average length of journey to the countryside was 27 km, but with almost one-third of the sample identifying walking as their means of transport, a significant part of the usage would be much more localised than the mean figure implies. Kay (1996) shows that

Table 5.1 Frequency of countryside visiting and the share of visits in the UK, 1990

Frequency of trips in any 4 weeks	% share of population	% share of visits	Category of visitor
9 or more trips	11	61	Regular and frequent users
5–8 trips	9	19	
2–4 trips	18	16	Casual users
1 trip	12	4	Occasional and infrequent users
No trips	50	0	

Source: Kay (1996).

much of the pressure of use upon the countryside is a consequence of the activities of a comparatively small section of the population who are regular and frequent users, often engaging in short-duration trips for simple purposes such as exercise or dog walking (Table 5.1). Curry (1994), Harrison (1991) and Sharpley (1996) all draw attention to the stabilisation through the late 1980s and early 1990s of the proportion of the population that visit the countryside, which suggests – as Kay implies – that continuing growth owes more to the increasing frequency with which a minority of users visit, rather than widening participation within the population at large.

Diversification of activities and locations

The growth in participation is intimately linked with processes of diversification. Butler *et al.* (1999) argue that up until the last two decades of the twentieth century, recreational and tourist activities in the countryside were closely related to the character of the setting. Typical activities (such as walking/rambling, picnicking, sightseeing, fishing, sailing or riding) could be loosely characterised as relaxing, passive, nostalgic, traditional, low-technology and essentially non-competitive. Whilst such activities remain at the heart of many recreational or tourist visits to country areas (see Table 5.2), other – rather different – activities have been added to the list. These typically embody opposing qualities: they are active, competitive, prestige or fashion-based, high-technology, modern recreations. Examples would include mountain biking, off-road driving, orienteering, survival games, hang gliding, parasailing, jet-ski boating, wind surfing and extreme sports such as bungee-jumping – new activities that infuse both recreation and developing areas of adventure tourism.

These changing patterns of demand often create a requirement for the development of more specific facilities and, occasionally, new resorts to cater for public tastes that are becoming more specialised and sophisticated. When these newer demands are married with a growth in participation in some of the more traditional recreational activities, significant new pressures are seen to emerge. Clark *et al.* (1994) itemise several areas of recent growth of recreation

Table 5.2 The continuing appeal of traditional countryside activities amongst visitors to national parks in the UK

Activity	% of visitors participating
Sightseeing by car	70
Sightseeing on foot	63
Eating/drinking out	55
Walking (less than 1 hour)	45
Walking (between 1 and 4 hours)	43
Shopping (for gifts and souvenirs)	38
Picnicking	36
Visiting historic sites	31
Relaxing/sunbathing	29
Hill walking	18
Cycling	6
Riding/pony trekking	3
Angling	3

Source: Based on information from Countryside Commission (1996).

and tourism in the British countryside that have created conflicts and pressures, including the growth of rural theme parks, the development of rural sports and the development of integrated holiday villages.

The growth of rural theme parks

Although the theme park has an extended history within the development of American leisure (with the original Disneyland at Anaheim, Los Angeles, opening in 1955), British theme parks are much more recent in origin. The majority date from after 1980 although in some cases, theme parks have been developed around older attractions, thereby illustrating impacts associated with changing popular tastes. The Chessington World of Adventure (Surrey, England), for example, was redeveloped from Chessington Zoo in 1987.

Whilst some of the parks are peri-urban in location (for example, Legoland at Windsor or Thorpe Park at Staines – both on the outer fringes of the London conurbation), the significant land requirements and the desirability of setting the rides and attractions into a pleasing, scenic context, have ensured that the majority of the main parks are actually rural in location – see Figure 5.7. Consequently, these new attractions have been responsible for drawing very significant numbers of visitors to rural areas. British theme parks attracted in excess of 12.6 million visitors in 1999 (Mintel, 2000c), with the largest – Alton Towers in Staffordshire – receiving around 2.65 million visitors annually. The site covers over 240 ha, although this is small in comparison with Disneyland Paris, where some 1,800 ha of rural land were used in the development, and where around 11 million visits are made each year.

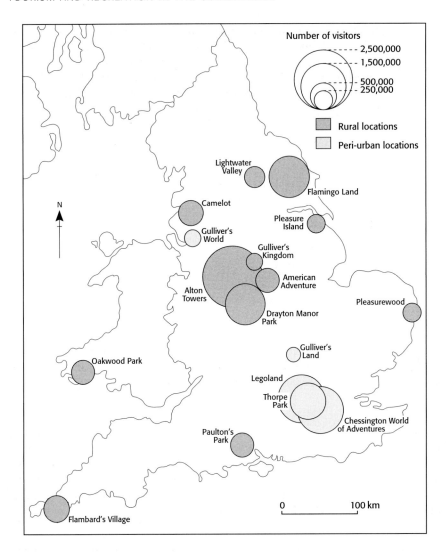

Figure 5.7 Locations and levels of visiting to major theme parks in England and Wales, 1999
Source: Based on information from Mintel (2000c).

The development of rural sports

As we have seen in the first part of this chapter, sport formed an early leisure use of the countryside and traditional rural sports – such as hunting, shooting, riding and, especially, fishing – continue to be popular. However, these established rural sports and outdoor pursuits have been widely supplemented by newer activities that present different demands. For example, McGarvey (1996) reports that in the area of water-based recreation in Britain, canoeing now attracts more than 1 million, motor cruising on inland waterways some 750,000 and water skiing over 400,000 participants a year. Increased affluence and technological developments in the design and production of equipment have made access to these sports and recreations more affordable, but have also extended the spatial patterns of participation. In activities such as canoeing, for instance, the development of craft built from high-impact plastics or carbon fibre have produced boats that are almost indestructible. This, in turn, has increased the number of rivers that can be tackled by 'white-water' enthusiasts.

Golf is another example of a growth sport that is exerting new pressures on the countryside. The significant land requirements in providing for golf ensures that developments are normally focused onto urban fringe and rural sites and as a consequence of popular demand, expansion in provision of rural courses in Britain in the early 1990s saw up to 36,000 ha of rural land a year being transferred to recreational use, mostly from agriculture.

Cycling is a further example of a popular recreational sport that is showing a resurgence in popularity. Interest in cycling has been restored through a combination of factors including:

- greater public awareness of health benefits related to exercise;
- the development of mountain bikes and of mountain biking as a fashionable outdoor recreation, especially amongst young adult groups;
- media exposure of cycling as a sport, especially prestige events such as the Tour de France.

Hence, the numbers of recreational and touring cyclists have risen, exerting new levels of demand and stimulating new patterns of provision in the countryside. The following case study illustrates a recent, major initiative to develop and promote cycle tourism in Europe and provides a picture of growth that contrasts with the earlier pattern of decline described in the case study of cycling in France used earlier in this chapter.

Case study: Cycle tourism in Europe: the development of EuroVelo

The proposed development of EuroVelo as an international network of cycle routes across Europe reflects wider concerns to produce sustainable general transport policies that are less dependent upon the car, as well as realising the more specific potential for cycle tourism to act as a catalyst to growth in rural areas across the continent. The scheme is being developed and funded by a wide variety of partner organisations and municipalities across the EU and beyond, under the guidance of a project management team. This is formed by the European Cyclists Federation, Sustrans (who are developing the National Cycle Network in the UK), and De Frie Fugle – a Danish organisation specialising in sustainable planning.

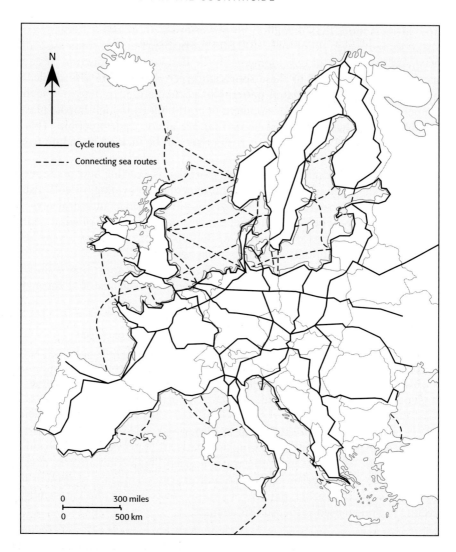

Figure 5.8 The proposed EuroVelo cycle network in Europe
Source: After Lumsdon (2000).

When fully developed (by 2010), EuroVelo will comprise some 61,709 km of cycleways (Figure 5.8). In some areas EuroVelo will incorporate existing cycleways (or routes already under development – such as the Sustrans Network in the UK), but elsewhere, the scheme will require significant provision of new routes. As far as possible, these will form dedicated cycle paths (especially where the network enters major towns or cities), although quiet rural lanes will also form key pathways. Each route will have a theme – such as the 'Pilgrims' Route', which follows the medieval pilgrimage trail from Trondheim (Norway) to Santiago de Compostela (Spain).

Table 5.3 Overnight stops by cycle tourists on part of the Danube Cycle Route (*Donauradweg*), Austria, 1994

Location	No. of overnight stops
Freinberg	4,114
Vichtenstein	25,766
Engelhartszell	26,265
Neuslift	24,663
Hofkirchen	8,086
Waldkirchen	9,253
Haibach ober der Donau	57,018
Aschbach	8,221
Feldkirchen	17,952
Ottensheim	12,225
Linz	458,930

Source: Lumsdon (2000).

The proponents of EuroVelo draw considerable encouragement from evidence of growing demand for cycle tourism in Europe. Data suggest that across Europe as a whole, dedicated cycle tourists account for up to 4 per cent of holiday trips, whilst in countries where traditions in cycling are stronger (for example, Denmark and Germany), levels are as high as 8 per cent. More significantly, perhaps, occasional cyclists and people who cycle as part of a holiday or same-day visit to the countryside, are much more numerous. Up to a quarter of Germans, for example, report some use of cycles for holidays or recreational trips. The Munsterland province of Germany already possesses over 10,000 km of cycleways which attracted an estimated 15 million cycle trips in 1995.

The economic impacts of the scheme are also judged to be potentially very valuable, especially at the local level. In areas where cycle tourism is already well developed, evidence shows significant levels of local spending in rural communities and small towns, especially on accommodation, food and drink. On the Danube Cycle Route (*Donauradweg*) in northern Austria, an estimated 1.2 million visitors currently spend an average of 66 euro per day (£40), and although trips may be no more than a few days in duration, local impacts are still significant. Table 5.3 illustrates how such spending might be distributed amongst smaller communities in the area between Freinberg and Linz, on the basis of overnight stops by touring cyclists.

Whilst the creation of the full EuroVelo network lies in the future, analysts are confident in the strength of the market for cycle tourism. Realising the full potential of EuroVelo will, however, require not only provision of routes that are suited to the needs of recreational or touring cyclists, but also effective signage and publicity, provision of supporting infrastructures (especially accommodation), and active promotion.

Source: Based on information from Lumsdon (2000).

The growth of integrated holiday villages

Forms of recreation and tourism based in holiday centres have been established for some time, with traditional holiday camps (now rebadged as 'holiday centres') operated by firms such as Butlins dating from the 1930s. These have been augmented by the more recent development of, first, holiday parks (which are generally based on a combination of mobile homes or chalets with on-site leisure and entertainment facilities) and second, integrated forest holiday villages belonging to companies such as Oasis and Center Parcs.

In Britain, this sector is currently thought to provide around 6 million holidays each year (Mintel, 1999b) with the largest share of the market held by the traditional holiday centres, especially Butlins. (These facilities are essentially coastal in location and are mainly concentrated in the vicinity of the older, urban resorts.) However, in rural recreation and tourism, it is the integrated forest village that is attracting most attention.

In this sector the formerly Dutch-owned company Center Parcs has been largely responsible for developing the concept. The first Center Parc holiday villages were established in the Netherlands, but subsequent development has seen 13 centres established in France, Belgium and the UK – all in rural locations (Gordon, 1998). The first British Center Parcs development was at Sherwood Forest (opened in 1987) with further villages located at Elveden Forest and Longleat (Wiltshire). Oasis has entered the market more recently, with its first centre outside Penrith (Cumbria) and a second scheme currently being developed in Kent.

The concept leans heavily upon the positive attractions of the natural environment to create a self-contained holiday centre, complemented by carefully managed artificial environments. Each site (which will typically extend up to 160 ha) is selected to provide high-quality, ecologically diverse landscapes of woodland, heath, glade, meadow, lakes and streams. These provide a background for popular outdoor recreations (such as walking, cycling, riding, golf and a range of water-based activities), whilst also creating a strong sense of seclusion within a country setting. Additionally, each centre offers indoor spaces that include a subtropical water world contained within a controlled environment for all-weather use, as well as a selection of restaurants, bars, shops and games facilities. Self-catering accommodation is provided for up to 3,500 people at a time (Gordon, 1997). In 1998, Center Parcs in Britain attracted more than 650,000 visitors, especially in family groups and primarily from the wealthier ABC1 sectors of the population.

These different developments have all contributed to recreational and tourism activity becoming more spatially diffuse. Although some forms of activity remain concentrated by the pattern of supply or opportunity, the development of a growing range of managed sites for rural recreation and tourism means that those opportunities are becoming more widely distributed. More significantly, however, it is the evident demand for public access to a wider, unmanaged (or more loosely managed) countryside that is responsible for a progressive widening of the geographical patterns of the countryside visitor.

Increasing commercialisation and commodification

Integral to the development of the widening range of rural recreational and tourism sites and attractions have been the processes of commercialisation and commodification. These are conspicuous within several of the examples given above (in particular, the growth of rural theme parks or the development of integrated holiday villages), but the impact is evident in the rising number of other attractions. Typical examples include restored steam railways, working farms, wildfowl centres, rural museums, craft centres, fish farms, collections of rare breeds, restored mills and a diverse range of heritage and other types of visitor centre.

Cloke (1993) suggests that the increased incidence of commercial attractions and a pay-as-you-enter countryside are consequences of shifts in the political climate (especially under the agenda of the New Right) towards privatisation and deregulation. This, it is argued, has opened new opportunities for a range of public and private agencies, as well as individual entrepreneurs, to develop the economic potential of recreation and tourism by creating new visitor-based attractions. However, the fact that people are increasingly willing to subscribe to a more commercially oriented countryside owes much to wider shifts in the culture of consumption (Urry, 1995), in which public susceptibility to buying 'packaged' recreational experiences has become more pronounced. Thus, the countryside has become both a context for the consumption of recreation and tourism, as well as a consumable entity in its own right (Ashworth and Voogd, 1994).

Underpinning this process of commodification is a dependence upon the existence of sets of popular images of the countryside. Hopkins (1999) argues that rural recreation and tourism attract participants by accentuating the sense of difference between town and countryside. This is achieved by a complex system of semiotics that evokes myths of rurality, simultaneously promoting and feeding off a nostalgia for rural life. Hence, for example, visitors to the English Lake District will do well to escape the nostalgic reconstructions of a vanished England that are encapsulated in the 'Peter Rabbit' stories of the Lakeland writer Beatrix Potter. Her anthropomorphised creations adorn souvenir tea towels and biscuit tins across the length and breadth of the national park, and in 1999 more than 225,000 people paid to enter the 'World of Beatrix Potter' exhibition at Windermere (Mintel, 2000c). Nor is this a uniquely British preoccupation. In North America, Bunce (1994: 134) notes that 'most of the commercialised countryside has been developed to serve an affluent consumer, for whom a weekend staying at a classy country inn and browsing around antique and craft shops has become very much an extension of fashionable lifestyle'.

The construction of images used in rural tourism has been studied by Cloke (1993) in the context of west Wales and Hopkins (1999) in southern Ontario, Canada. Both authors found a marked degree of similarity in the approaches to the use of imagery in promotional literature for tourism in the two study areas. In particular, the qualities of the landscape and its wildlife were widely invoked to instil a sense of nature and associated qualities of tranquillity, escape and peace, whilst history, images of family and an emphasis upon local production (such as foods and craftwork) were used to reinforce ideas of tradition.

In many cases, the promotional literature studied by both Cloke and Hopkins revealed the increasing self-inclusivity of commercial attractions in modern rural areas. Sites were widely marketed as places to spend a day and to achieve this aim the main attractions were typically supported by opportunities to visit souvenir and craft shops, eat in the restaurant or at the picnic areas, walk the grounds, enjoy the scenery, feed the animals, ride the vintage vehicles and so on. Figure 5.9 illustrates an example of such an inclusive site – Amerton Working Farm, in the English Midlands. This complex has been progressively developed as a local and regional attraction over a period of more than 20

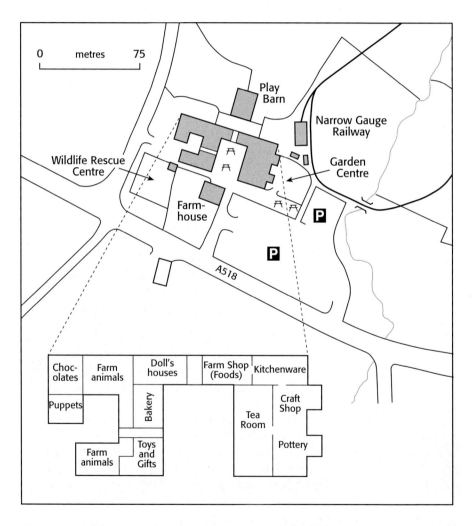

Figure 5.9 Visitor attractions at Amerton Farm, Staffordshire

years, and now draws more than 185,000 visitors annually (Mintel, 2000c). Its success is based upon:

- its ability to encapsulate essential qualities of a day out in the countryside with its combination of landscape, farming and the purchase or consumption of rural produce, such as food or handicrafts;
- the quality of the goods and services that it retails (which are aimed squarely at the more affluent and discerning ABC1 social groups);
- the all-inclusive nature of the site which offers walks, rides on a narrow-gauge steam train, a range of craft workshops, farm animals and a wildlife sanctuary, retail areas selling foodstuffs, kitchenware, high-quality gifts, a garden centre, a restaurant, toilets, and an extensive area of free parking.

Integration of recreation and tourism into the rural economy

The wider integration of recreation and tourism into the rural economy (as exemplified at Amerton Farm) has been a part of a much broader pattern of change in rural areas that occurred across the developed world over the last quarter of the twentieth century (see Figure 5.10). Jenkins *et al.* (1999: 52) comment that 'despite a lingering image of stability and tranquillity, and a traditional way of life, the reality of the rural sector in all developed countries

Figure 5.10 Integration of recreation and tourism into the productive rural economy – facility development at Bewl Bridge reservoir, Sussex

is of rapid and disruptive change'. Impacts of globalisation have seen production processes increasingly integrated across national boundaries and subject to the control or influence of multinational corporations, whilst at the regional and local levels, country areas have been subject to a host of changes, many of them conflicting in their nature or effects. Thus:

- Although rural populations have risen, so they have become more urbanised and/or elderly – through the attraction of the countryside as a place of residence for urban workers and, eventually, for retirement.
- Whilst traditional rural industries such as farming have become more productive through technological advances and integration into a globalised economy, overproduction has driven down prices and profits, creating hardship, contraction in the farm workforce and unemployment.
- Whilst closer links between town and countryside have been promoted through better personal transportation and communications, local rural services (such as shops, schools and health care) have tended to disappear, either through an inability to compete within an urban-dominated marketplace, or through processes of rationalising public services.

These changes have impacted markedly upon both the rural economy and the social coherence of rural areas which – when combined with heightened public sensibilities to the need to conserve and maintain the natural and cultural heritage of rural areas – has created a new policy agenda for the countryside.

Within this new agenda, provision for recreation and tourism has occupied an increasingly centralised position, reflecting both the active growth of demand as well as the perceived opportunities for local rural economies to diversify traditional livelihoods and develop new sources of income. Although it is well understood that recreation and tourism are not necessarily a panacea for all the problems created through restructuring of the countryside (see, for example, Hall and Page, 1999 who document a range of potential difficulties associated with developing rural tourism), active expansion of provision through integration with other rural activities has been a conspicuous (and often successful) development. In addition to specific sectoral benefits, research in the UK also suggests that a range of wider community benefits may accrue through development of rural tourism and recreation (RDC, 1996). These include:

- enhancement in the viability of rural retailing, including non-tourism businesses;
- creation of employment for local tradespeople, such as builders;
- support for key services such as local transport, banks and post offices;
- contributions to social and community life, for example through support for local festivals and events;
- environmental benefits through visual enhancement, reuse of redundant buildings and encouragement of better upkeep of property.

According to the Countryside Agency (1999), by the mid-1990s, total spending by visitors to the UK countryside stood at an estimated £9 billion, directly supporting over 350,000 jobs.

To illustrate the trend towards integration, the final section of this chapter considers some of the ways in which tourism and recreation have been developed within the primary traditional rural sector – agriculture.

The integration of recreation and tourism with agriculture

Historically, agriculture was a sector that was shaped by an economic imperative that saw provision for wider amenity and leisure interests pushed to the margins of operation, or – more typically – excluded altogether. However, since the mid-1980s, agricultural overproduction within many parts of the developed world – and especially in the EU – has prompted significant changes in policy and practice. Reform of the EU Structural Funds in 1988 and of the Common Agricultural Policy in 1992 has led to a dismantling of many of the production-centred support mechanisms for farming and a new emphasis upon encouraging diversification of enterprises that are no longer capable of standing as production units alone (Slee *et al.*, 1997). Additionally in Europe, regional initiatives in the EU's LEADER programme have included many ventures aimed at fostering rural tourism as a means to replace declining agricultural incomes and tackle the problems that surround contracting employment opportunities (see Hall and Page, 1999: 210). Across much of Europe agricultural employment has collapsed. Between 1980 and 1996, the farming workforces in Italy and Germany contracted by more than 60 per cent, in Greece agricultural employment fell by 50 per cent and in the UK by 17 per cent (Edmunds, 1999). In the UK 60,500 jobs were lost in farming between 1987 and 1997 (Countryside Agency, 1999), signalling very clearly the need for an alternative strategy for farming within the rural economy.

Estimates of the extent to which farms have diversified into sectors such as recreation and tourism vary, although the composite picture points to the overall significance of the trend. Denman (1994) suggests that in Britain, nearly a quarter of farms now contain some elements of provision for recreation and tourism. With over 144,000 holdings in the UK (Countryside Agency, 1999), this provides an aggregate figure in excess of 30,000 farms with some interest in leisure provision, although much of it will be small-scale and incidental. Similarly high figures are also found elsewhere. In Austria, for example, where traditions of receiving tourists on farms date back more than a century (Frater, 1983), half of all tourist bed spaces are to be found on farms (Busby and Rendle, 2000). However, in the USA and Canada participation rates are much lower (Fennell and Weaver, 1997). A study of farm tourism in Saskatchewan (Weaver and Fennell, 1997), for instance, found that only 0.1 per cent of the province's farmers had diversified into farm tourism. Issues of accessibility, the small size of the provincial population and the limitations on the length of the season as a consequence of the harsh Saskatchewan winters were all seen as influential in limiting development in this case.

This example reminds us that the opportunity to diversify into recreation and tourism does not occur uniformly. McNally (2001) emphasises spatial variations in opportunity that mirror variations in the pattern of tourism itself. A study by Evans and Ilbery (1992) identified nearly 6,000 farms with accommodation in England and Wales, with the majority coinciding with key tourism regions such as the South West, the Lake District, the Welsh Borders, north Yorkshire and coastal areas in the South East. Variation also occurs according to the scale and nature of the farm enterprise. Larger, crop-based farms are better placed to develop other forms of activity than smaller, livestock farms (McNally, 2001). Small enterprises also tend to suffer from the fact that the additional income that these units are able to generate through recreation and tourism is modest and makes only a marginal contribution to the overall profitability of the enterprise (Bowler *et al.*, 1996). As a consequence, the long-term stability of diversified enterprises has sometimes proven to be uncertain, and the similarity between rates of entry and rates of exit in diversification schemes suggests quite high rates of failure (McNally, 2001).

The nature of diversification is itself also variable. Some forms of diversification do not involve recreation and tourism at all and concentrate instead upon provision of agricultural services, subcontracting, or the leasing of farm buildings to other enterprises. However, within the recreation and tourism sectors, an analysis by McNally (2001) of 2,979 farm businesses in the UK Farm Business Survey found that provision of accommodation for tourists was the most commonplace form of leisure-based diversification, either through traditional farmhouse bed and breakfast or through new construction or conversion of farm buildings to self-catering accommodation units. Usage of such accommodation is characteristically of short duration, with up to two-thirds of visitors to UK farms staying for less than four nights and nearly a third for just one night (Turner and Davies, 1995). Nevertheless, accommodation services contribute the largest share of an estimated £70 million generated annually through farm tourism in Britain (Busby and Rendle, 2000). Visitors are drawn by the attractiveness of the rural setting, peace and quiet and value for money, as well as opportunities to experience, first-hand, a 'traditional' rural way of life (MAFF, 1994; Turner and Davies, 1995).

In contrast, the recreational sectors in farm-based leisure are less significant in their scale, frequency or contribution towards farm incomes, but are very much more diverse in nature. Table 5.4 reflects this diversity within a typological framework adapted from work conducted by Ilbery (1989) and which draws a key distinction between resource-based recreational activities and enterprises developed as day visitor attractions. Ilbery's (1991) work also makes an important spatial distinction in emphasising the significance of many of these opportunities to farmers working in urban fringe areas and therefore close to sizeable local populations. In fringe areas, diversification into tourism

Table 5.4 A typological framework for farm-based recreation

Resource-based activity	Enterprise-based activity
Horses or ponies	*Informal recreation*
Riding and trekking	Picnicking
Eventing	Walks and trails (including nature walks)
Livery facilities	Car parking
Grazing	Country parks
	Bird watching
Water	
Fishing	*Catering and retailing*
Canoeing	Teas
Boating	Farm produce (shop or roadside)
	Pick-your-own
Shooting	Crafts
Game birds	
Clay pigeons	*Education*
Water fowl	Open days
Archery	School visits
Rifle and pistol	Demonstrations
Sports	*Events*
Golf (including driving ranges)	Shows
Motorcycle scrambling	Ploughing matches
	Gymkhanas

Source: Adapted from Ilbery (1989).

is often inappropriate, but provision for local recreation is much more worthwhile.

The preceding discussion largely reflects the recent experiences of farm-based tourism and recreation in Britain. As an alternative perspective, the final case study in this chapter examines farm tourism in New Zealand.

Case study: Farm tourism in New Zealand

As in the UK, the agricultural landscape of New Zealand constitutes a primary attraction for rural tourists and recreational visitors. Unlike the UK, however, the development of the farm tourist sector has occurred on only a limited basis so far. This is due to the relatively small size of the domestic tourism market in New Zealand (the population of which is only 3.5 million) and the modest scale of international tourism to one of the most isolated destination countries in the world. International arrivals in 1996 numbered only 1.5 million, although this represents a significant increase in percentage terms upon levels of visiting at the start of the 1980s, which stood at less than half a million visits per annum.

The development of farm tourism has been stimulated by factors that are similar to those experienced in Britain and Europe. Traditional farming subsidies were widely abolished after 1984, rendering New Zealand farmers far more vulnerable to the fluctuations and uncertainties of global agricultural markets. However, the coincidence of subsidy removal with a significant expansion in international tourism helped to create new demands and new opportunities for farm tourism. Hence, the provision of farm tourism in New Zealand is characterised by its comparative recency, with an estimated 82 per cent of enterprises that are currently operating having opened since 1985.

In the UK, international tourists are believed to comprise only a small percentage of visitors on farms (MAFF, 1994). In contrast, in New Zealand, nearly three-quarters of the demand comes from international visitors. Farm stays only account for some 7 per cent of international tourist bed-nights, but if people who stay with friends and relatives are discounted, that share rises to 11 per cent and provides a clearer picture of both the potential and the actual value of farm tourism.

The scale of tourism provision is characteristically small, with 55 per cent of operators having four or fewer beds for use by visitors. Typically these are provided in existing accommodation – usually spare rooms within the farmhouse – rather than through new construction or conversion of premises, as is commonplace in the UK. The length of stay, as in the UK, is normally short, principally because the international tourists that dominate the farm-stay sector in New Zealand tend to be touring the country with a view to seeing as many places as possible. However, because of the use of surplus accommodation (which reduces the need for capital investment) and the high level of dependence upon the farm family to provide labour, costs are reduced to the point at which the smallest enterprises can be profitable, even where the individual length of stay is brief or where the tourist season is relatively short – as on the South Island. Interestingly, although financial gain and stability of income are a primary motive for running farm-based tourism on nearly a third of farms in New Zealand, social contacts between farmers and visitors is rated more highly as a reason for providing for tourists. Many New Zealand farmers work within small, isolated communities, so the opportunities for encounters with others as tourists is valued in a way that would be unusual in, say, a British context.

Source: Based on information from Oppermann (1999).

Conclusion

From the preceding discussions, several key ideas emerge. Perhaps most importantly, first, we should note the increasing significance of recreation and tourism within what some writers now choose to label as a 'post-productive' countryside. As less emphasis is placed upon the use of rural land for agriculture, forestry, mineral extraction or water catchment, so the amenity role of the countryside and its value as a set of leisure sites tend to be enhanced.

However, although of increasing significance, those leisurely uses remain highly uneven. There is a selectivity in terms of who uses the countryside and, additionally, a marked unevenness in spatial and temporal patterns. As a result of the spatial and temporal patterning, in particular, many recreational and tourist visitors routinely converge on the same sites, at the same times – producing a blend of tourist and recreational practices at popular locations. We have also seen how this mutual interdependence and association are reflected in activity, especially through the growing trend towards forms of tourism based around popular countryside recreations, as well as in some areas of provision.

Finally, the chapter shows through its historical review, how the closer relationship between rural recreation and tourism – whilst becoming a more conspicuous element in the (post) modern era – actually has visible origins in the earliest phases of leisurely use of the countryside. This suggests, once again, that whilst ideas of de-differentiated forms of leisure, recreation and tourism may indeed be 'emblematic of postmodernity' (Crouch, 1999: 1), the process is far from being confined to the present.

Questions

1. What were the factors that shaped patterns of rural recreation and tourism before 1939?
2. To what extent is the appeal of the countryside a selective one, and what reasons may be advanced to explain patterns of selectivity that may be detectable?
3. Explain what you understand by the term 'commodification' and show how the concept is being applied in the development of new attractions in the countryside.
4. Why has integration of recreation and tourism become a prominent theme in contemporary rural policy?

Further reading

Historical patterns of rural recreation and tourism are well documented in Towner, J. (1996) *An Historical Geography of Recreation and Tourism in the Western World, 1540–1940*, Chichester: John Wiley, whilst a fascinating cross-cultural perspective is provided in Bunce, M. (1994) *The Countryside Ideal: Anglo-American images of landscape*, London: Routledge.

For succinct discussions of general patterns of provision and use in countryside recreation, see Glyptis, S.A. (1991) *Countryside Recreation*, Harlow: Longman; Harrison, C. (1991) *Countryside Recreation in a Changing Society*, London: TMS Partnership; and appropriate chapters in Hall, C.M. and Page, S.J. (1999) *The Geography of Tourism and Recreation: Environment, place and space*, London: Routledge.

Contemporary themes and issues are also covered in an interesting collection of recent essays in Butler, R., Hall, C.M. and Jenkins, J.M. (eds) (1999) *Tourism and Recreation in Rural Areas*, Chichester: John Wiley.

Tourism and recreation: issues and policy approaches

CHAPTER 6

Introduction

The dominant theme in the previous four chapters has been the interplay between recreation and tourist practice – as exemplified across a range of spatial and temporal scales, and within contrasting urban and rural contexts. To conclude the book, however, this final chapter deploys a shift in emphasis towards the consideration of policy, as it affects – and is affected by – recreation and tourism. In particular, it offers a broadly based discussion of some of the key issues or problems that are raised by the development of recreation and tourism (especially as a parallel process), and some of the approaches that have emerged in the quest for solutions to the issues that are raised.

The discussion is arranged within three linked sections. First, a range of key issues are identified. These centre on:

- the constraints on development that are posed by problems such as land availability, landownership and deficiencies in associated policy;
- the conflicts and impacts that are created by pressures of demand for recreational and tourist resources.

Second, the nature of some of the primary avenues of response to these issues is explored, especially in terms of the development of new policy frameworks for recreation and tourism and the associated emergence of appropriate planning and management techniques. Finally, the interplay between issues and responses is illustrated by reference to two contrasting areas of development in which recreation and tourism are often brought together under a common policy umbrella:

1. Urban renewal programmes.
2. The widening of public access to the countryside.

In view of the difficulties of generalising from a range of contrasting national policy contexts, most of the discussion is centred upon the British experience, with a limited number of comparative references to other places. There are, however, wider lessons that may be learned from the British examples, and the progressive integration of Britain into EU policy structures and institutions also ensures that policy and practice in the UK does now reflect issues that resonate within a wider geographical arena.

The approach adopted in this chapter is, of necessity, selective. There is, for example, no discussion of the extensive policy areas associated with tourism planning, but this theme is already well covered in the literature (see, as examples: Gunn, 1988; Inskeep, 1991 and a recent companion volume in the *Themes in Tourism* series: Hall, 2000). But it does attempt to consider issues and approaches that have a broad relevance to contemporary development of recreation and tourism opportunities, in both urban and rural situations.

Recreation and tourism: key policy issues

There is a diversity of policy issues that affect recreation and tourism. These include a series of essentially practical questions relating to issues such as planning (e.g. location, accessibility and design of facilities) as well as management – especially in situations where recreational or tourist activity needs to be harmonised with other resource uses. For purposes of this present discussion, however, two key areas that are perhaps more fundamental than planning and management policies are explored in fuller detail. These relate, first, to the basic constraints upon the supply of recreational and tourist opportunity, and second, to the impacts and conflicts that recreational and tourist use of resources may generate.

Constraints

In the UK, policy directed towards the development of recreation and tourism is (and has been) generally affected by three, interrelated areas of constraint: availability of land; patterns of ownership; and deficiencies in policy, especially prior to 1990.

As resource-based activities, tourism and recreation opportunities are shaped fundamentally by the availability of appropriate resource areas – whether land- or water-based. In most, though not all, situations this tends to create problems of competition with other potential users of the same space. In the context of rural recreation and tourism, Jenkins and Prin (1999: 182) observe that 'other uses include primary production (e.g. agricultural, aquatic, horticultural, pastoral and timber production), conservation or preservation of the natural and built environment (e.g. national parks and wilderness areas), and transport and communications networks'. Similarly, in towns and cities, allocations of space to recreation and tourism have to be assessed against the competing needs of housing, retailing, industry, administrative functions, office and commercial activity, distributional services and transportation.

For a variety of reasons, recreation and tourism have seldom been accorded priority over any of these sectors and have often competed for land and other resource allocations from a position of relative weakness within many areas of policy discussion. This subordinated position has arisen through:

- the traditional strategic significance attached to sectors such as agriculture or forestry;
- the self-evident importance of social and economic policy in areas such as housing, public health and employment;
- the fact that recreation (although not tourism) has been viewed as producing only minor economic benefits;
- the fact that the recognition that is now being accorded to recreation and tourism (as defining features of personal and collective lifestyles) is a relatively recent phenomenon.

Consequently, in congested areas such as the UK, allocations of land and water space that are dedicated to recreation and tourist use have been minimal although, as will be seen later, in countries where land is more widely available – for example, the USA – rather different patterns have emerged. Dedicated leisure sites are an established feature of the contemporary urban environment (in parks, gardens, recreation grounds, sports fields, golf courses and a range of entertainment facilities), although their physical extent is often quite limited in relation to the total land area of a typical town or city. In the countryside, however, the dedicated sites that do exist (in, for example, country parks, visitor centres, theme parks, caravan and camping sites, or rural tourist attractions) comprise only a tiny fraction of the overall rural space. Instead, most demand for rural recreation and tourism is met from land and water spaces that have other (primary) functions – a pattern that places a premium upon the ability to manage land and water resources for multiple use.

The issue of availability of land is linked directly to a second area of constraint, the patterns of landownership and the restrictions that ownership exerts upon public access to private land for recreation or tourism. Jenkins and Prin (1999: 182) again state the problem concisely when they write that 'land ownership and the exercise of landownership rights are critical elements in the supply of tourism and recreation opportunities, because access to land and water ... is generally contingent upon legislation, public policy interpretations, and landholder/management attitudes'.

Issues of ownership are especially evident within rural recreation and tourism where they may be seen to be influential in several contrasting ways. There is, for example, a clear link between landownership and the exercise of political power and influence. Blunden and Curry (1990) show how the popular struggle to secure wider access to the British countryside in the period between 1920 and 1950 was repeatedly confounded by the capacity of landowners (both inside and outside Parliament) to affect decisions in favour of the broad protection of conventional proprietorial rights of ownership. In so doing, this created inevitable tensions since, as Cherry and Rogers (1996) observe, over the course of the twentieth century the urban community formed their own collective sense of 'ownership' of the countryside. 'In 1914,' they write, 'the only claims to countryside use and access which were recognized were those which were vested in ownership. Eighty years on that is no longer the case. The evidence is overwhelmingly that the ordinary urban public feel that they have a stake in "their" countryside' (Cherry and Rogers, 1996: 203).

More directly, however, recreational and tourist visitors to rural areas encounter the influence of ownership through their routine exclusion from private land. According to Shoard (1987), 87 per cent of the land in the UK is privately owned and even where land is in public ownership – such as within the holdings of the Forest Authority – public rights of access for purposes of recreation and tourism are provided on only a selective basis (Glyptis, 1991b). Similarly, within the national parks of England and Wales – which function as key recreational and tourist environments and draw over 75 million visitors each year – the levels of private holding of land rise to as high as 96 per cent (in the Yorkshire Dales National Park) (Countryside Commission, 1993a). Table 6.1 shows that this level is exceptional, but even so, the general dominance of private land in these areas raises relevant questions over the extent to which British 'national' parks are national in any meaningful sense of public ownership and associated access.

Not only is the level of private landholding in the UK historically high, recent trends have tended to reinforce its significance. Dwyer and Hodge (1996) argue that over the first half of the twentieth century the progressive dismantling of rural estates saw a significant increase in the level of owner occupation of agricultural land which, in itself, created new levels of restriction. More recently, an opposing trend – shaped by processes of rationalisation and farm amalgamation – has seen ownership of rural land reconcentrated into the hands of large-scale institutional owners, many of which have no direct connection with rural areas (such as finance and insurance companies). In a similar vein, Curry (1994) draws attention to the effects of policy shifts in the 1980s and early 1990s in favour of privatisation of public utilities and agencies. For example, the Forestry Act (1981) provided for the sale of Forestry Commission land to private landowners, whilst the Water Act (1990) recast the formerly public water authorities into new private companies, thereby increasing the threat of restrictions on access to water space for recreational or tourist users.

This links to the wider constraining influence of much of the policy formulation in recreation and tourism in the second part of the twentieth century. Here the 1949 National Parks and Access to the Countryside Act was particularly influential in creating an ambivalent attitude towards recreation and tourism that persisted for several decades. On the one hand, the Act offered new levels

Table 6.1 Patterns of landownership in national parks in England and Wales, 1990

	Percentage of land held by each owner									
	Brecon Beacons	Norfolk Broads	Dartmoor	Exmoor	Lake District	Northumberland	N. York Moors	Peak District	Pembroke coast	Snowdonia	Yorkshire Dales
Private	69.6	90.8	57.3	79.1	58.9	56.4	79.9	72.3	85.7	69.9	96.2
Forestry Commission	8.0	0.2	1.8	1.8	5.9	18.9	16.6	0.5	1.3	15.8	0.0
Ministry of Defence	0.1	0.0	14.0	0.0	0.2	22.6	0.5	0.3	4.6	0.0	0.3
Water companies	4.0	1.5	3.8	0.6	6.9	1.2	0.1	13.0	0.0	0.9	0.3
National Trust	3.5	3.0	3.7	10.1	24.2	0.7	1.2	9.6	4.2	8.9	2.5
English Nature/CCW	0.8	4.0	0.3	0.0	0.0	0.0	0.0	0.1	0.5	1.7	0.4
National Park Authority	13.0	0.5	1.4	4.4	3.9	0.2	0.6	4.2	2.3	1.2	0.1
Others	1.0	0.0	0.0	4.0	0.0	0.0	1.1	0.0	1.4	1.6	0.2

Source: Countryside Commission (1993a).

of provision and facilitation of recreation, whilst on the other, it established a primacy for conservation goals over those of recreation and reasserted the proprietorial rights of landowners (Parker and Ravenscroft, 2000).

It is arguable that subsequent legislation (up to the Countryside and Rights of Way Act, 2000) essentially perpetuated, rather than readdressed, the restrictive policy view of recreation and tourism promulgated by the 1949 Act. Both Harrison (1991) and Curry (1994) argue that the 1968 Countryside Act was shaped by (unfounded) fears over an explosion of recreation and tourism in the countryside, and whilst significant new initiatives were brought forward at this time (in the formation of country parks, designated picnic sites and a new emphasis upon long distance paths), the policy motivation was essentially shaped around principles of containment rather than widening access, in either a geographical or a social sense. Thus, for example, the newly formed Countryside Commission was originally given no remit to *promote* countryside recreation and tourism (Curry, 1994), and it was not until the publication of the 'Recreation 2000' strategy in 1987 (Countryside Commission, 1987) that clear signs emerged of a realignment in policy towards encouragement of participation and recognition of the previously ignored truth that countryside access was, in social terms, an exclusive practice (Harrison, 1991).

Significant criticism has also been directed towards the constrained manner in which recreation and tourist policy has been organised at governmental levels and within some of the key pathways for policy implementation – such as the statutory land planning procedures. Curry (1994), for example, notes the extended periods in which there has been no significant government activity on recreation and tourism (in the form of White Papers, Select Committee Reports, Parliamentary Acts, etc.) and adds that for much of the second half of the twentieth century, the organisational structure lacked stability. Patterns of ministerial responsibility for recreational and tourist issues were 'complex and cumbersome' (Curry, 1994: 40). Additionally, the statutory planning system has afforded inconsistent attention to recreation and tourism and has had to endure conflicting advice in the various governmental circulars and, more recently, Planning Policy Guidance that affect recreation and tourism development. As an industry with a growing importance in economic development strategy, tourism has been considered rather more fully than recreation – at both the national and the regional level (for example, in the Tourism Development Action Plans that have helped to regenerate declining urban seaside resorts such as Torquay). In neither case, however, has policy accorded a significant degree of priority to these two areas of leisure and this must be considered as a limitation.

Conflicts and impacts

The combination of a relative shortage of dedicated space for recreation and tourism, the high incidence of private ownership of land and a policy context that has tended to exhibit limitations in both philosophy and practice, has set up several conflicting tensions between recreational and non-recreational use. These tensions focus, in particular, upon:

- the relationship between different categories of recreational or tourist users, and between recreation or tourism and other activity;
- the relationship between recreation or tourism and conservation, especially in the countryside;
- the impacts that recreational or tourist activity create.

Relationships between recreation or tourist user groups have long been recognised as a problem area and a source of potential conflict wherever competing activities lay claim to the same space. Patmore (1983: 218) captures the essence of the problem in writing that 'too many activities may overcrowd the space; differing activities may interfere physically with each other; the aural or visual intrusion of one activity may seriously erode the enjoyment of the other'. This emphasises the need for resource managers to develop patterns of use that allow for integration of activities in ways that minimise the scope for conflict between user groups. (Examples of common approaches are provided later in this chapter.)

More fundamental than user conflicts, perhaps, are the problems of incompatibility between recreation or tourism and other land uses. In countries such as the USA these are minimised by the ability to allocate extensive tracts of land for primary use by recreationalists or tourists, but in the congested spaces of the UK, leisure users are often obliged to pursue activities on land or water that has other functions – for example, agricultural land. Here, the exercise of rights of privacy by landowners that were discussed above reflects the simple truth that the presence of visitors increases the risk of damage to property and/or a reduction in the productive capacity of the resource. Nor is this a uniquely British problem. A study by Jenkins and Prin (1999) of landholder attitudes in New South Wales (Australia) elicited a familiar list of problems for landowners that were generated by visitors and which were used by the farmers to justify the widespread exclusion of recreational visitors from their land. These included: damage to crops; disturbance of livestock (especially by dogs); failure to shut gates; littering; vandalism; indiscriminate shooting; and the lighting of fires.

Conflict between recreational or tourist activity and wider goals of conservation is similarly a familiar and well-established problem. In the UK, conservation policy – in both the urban and rural spheres – has been closely linked with recreation and tourism development, primarily because of the incidental positive influence that many areas of conservation practice exert upon the provision of recreational or tourist facilities and opportunities. For example, the protection of special landscapes and their wildlife; the conservation of historic buildings; the restoration of sites, buildings or artefacts that are now deemed valuable as 'heritage' – are all capable of creating attractions that draw leisure visits (see, *inter alia*: Fowler, 1992; Walsh, 1992; Larkham, 1996; Graham use, the term 'impact' tends to invite a negative reading, since impacts are often damaging in their consequences. In many leisure situations, negative impacts are indeed evident and there is a substantial literature that has documented detrimental change. This will include the:

- physical effects of recreational and (especially) tourist development – such as the construction of hotels or new transportation facilities; the visual pollution of landscapes with poorly designed caravan sites and holiday villages; or the transfer of land to non-productive leisure uses (such as golf courses);
- environmental effects – such as destruction of habitats, erosion and trampling of soils and vegetation by visitors at popular sites, or the pollution of air, ground or water by litter and waste, through vehicle emissions or perhaps the spillage of fuel from powered boats into water bodies;
- social and cultural effects – such as induced change in local customs, systems of beliefs and behaviours; or in the use of language;
- economic effects – such as inflation in local land values; change in local patterns of employment; or the creation of seasonality in patterns of income and earnings.

Equally, however, there may be positive impacts as recreation and tourism can stimulate a range of environmental improvement programmes; may bring investment and new sources of income or employment; may revitalise declining communities and provide stimuli to social change that is beneficial. (For a full discussion of recreation and tourism impacts see, amongst many: Mathieson and Wall, 1982; Murphy, 1985; Ryan, 1991; Hunter and Green, 1995.)

In many situations, impacts occur not in isolation but in combination and in ways that amplify their influence. An illustration of the composite effect of impacts – and one which places the emphasis squarely upon beneficial rather than detrimental change – is provided in Figure 6.1. This example relates to urban facilities and proposes a range of likely effects associated with the development of urban heritage and museum-based projects in the northern English city of Liverpool (Couch and Farr, 2000). In this example a range of positive influences are anticipated, including: physical enhancement of the urban environment; a strengthening of local communal identity; stimulation of local cultural life; and a range of economic benefits associated with project development and the spending of both local and tourist visitors in the sites in question.

This model offers no distinction between the effects of recreation and tourism, but tends to assume – as is commonplace in many discussions of policy areas – that the impact of these activities is essentially the same. In practice, however, the two sectors are associated with rather different effects. This is not so apparent in the areas of environmental change – where tourists and recreationalists engaged on the same activities will induce broadly similar impacts – but is certainly evident in areas of physical, economic and social change.

RECREATION AND TOURISM: KEY POLICY ISSUES   165

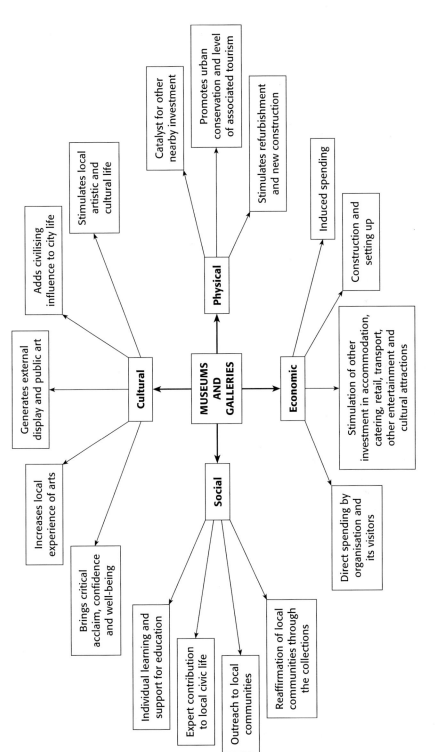

Figure 6.1 Positive impacts associated with heritage developments in Liverpool
Source: After Couch and Farr (2000).

As a generalisation, it is normally the impacts of tourists that exert the greater influence. This is due to several areas of difference. For example, first, the organisational basis of tourism (in which staying visitors are brought into an area from outside) increases the prospect for social or cultural impacts wherever there exist observable differences in social or cultural practice between residents and visitors. This applies only rarely in recreational contexts since, as an essentially local practice, participants are not displaced from their normal social or cultural milieu. In tourism, and especially in international tourism, the scope for more pronounced difference between host and guest, increases the prospect of a sociocultural impact through processes of acculturation or via the demonstration effect (see Williams, 1998: 152–4 for a concise explanation of these concepts).

Second, the more pronounced levels of economic activity that tourism tends to create will naturally produce stronger economic impacts. These may occur through the attraction of inward investment to provide tourism facilities; through the creation of a multiplier effect in local economies induced through patterns of tourist spending in key tourism sectors such as accommodation; or through the creation of employment within the tourism industry. Many areas of recreation produce no discernible economic effect since there is often no measurable financial exchange, but this is seldom true in tourism.

Third, the demands for particular types of facility to support tourism, especially accommodation but also other services such as transport and information, will prompt a range of physical impacts that would be much reduced (or absent altogether) if patterns were shaped purely by local recreation. Enhanced levels of provision of accommodation, restaurants, entertainment, physical attractions and local transportation are all likely consequences of tourism development. Some of these may be encouraged by recreational activity as well, but rarely to the same degree.

Recreation and tourism: responses to policy issues

The discussion in the first part of the chapter has established a range of key issues that policy for recreation and tourism needs to recognise and address. These centre upon a need to meet increased levels of demand with appropriate patterns of supply that permit recreation and tourism to coexist with other activities in ways that minimise detrimental impacts and potential conflicts of interest, whilst maximising potential benefits. What types of policy response to these issues have become evident? Two related areas appear to be especially influential:

1 The development of new policy frameworks that have allowed for a more formalised pattern of recognition of the needs and interests of recreation and tourism.

2 The development of management techniques that have enabled the wider accommodation of recreation and tourism alongside other sectoral activities.

Development of policy frameworks

The first section of this chapter has shown that in the UK, at least, recreation and tourism policy has followed a fine line between acknowledging a need for a widening range of leisure opportunities, whilst attempting to manage usage in ways that were essentially restrictive, particularly in the countryside. From the late 1980s onwards, however, policy was progressively restated so that:

- greater recognition has been accorded to the need for proactive development of recreational opportunity alongside other areas of resource development – often through processes of negotiation and the formation of public–private sector partnerships;
- the considerable potential of tourism as a force for economic regeneration in both urban and rural areas has been formally acknowledged.

Clark *et al*. (1994: 24) comment that 'overall, there is now an apparent official wish to develop leisure and tourism opportunities, locally and nationally, in a way that will pre-empt conflict between major public interests'.

In working towards this aim, several related changes have occurred. First, there is evidence of a greater level of complementarity between the consideration of recreational or tourist interests and the wider policy or management discussions of land use, resource development and environmental management. Second, there has been an increase in the designation of land or water that includes dedicated recreational or tourist space. Third, some policy areas have developed a more holistic approach to problem-solving, in which tourism or recreation have become key components in integrated programmes of managed change.

Resource development and environmental policy

Recreation and tourism have emerged as clear, although often indirect, beneficiaries of an evolving agenda that is centred around regional development and, especially, the sustainable management of resources and their associated environments. The EU, in particular, has become influential within its own boundaries in shaping new patterns of regional development, but in adopting and arguing for a progressive stance on environmental policy, Europe has also become a major voice in global discussions of problems such as atmospheric warming and large-scale pollution.

This represents a significant departure from the rather hesitant initiatives of the First Environmental Action Programme (EAP) of 1973, which spoke in only general terms of the enhancement of the quality of life for European citizens. The 1980s, in contrast, were an important period of change: first with the creation of the Environment Directorate-General (DGXI); then with the Third EAP (which was the first to offer a coherent statement of policies and aims for environmental improvement); and finally with the Single European Act (1987) which formalised and made explicit a strong EU involvement in the environmental field (Lowe and Ward, 1998a). Environmental policy was the fastest growing area of EU policy in the 1980s, reflecting increasing politicisation and public awareness of environmental issues and a realisation that environmental problems were becoming globalised and required a more integrated, harmonised response (Hashimoto, 1999).

Until recently, recreation and tourism seldom featured in discussions of EU environmental policy, let alone as primary topics. However, more recently the emphasis has begun to change. The Fifth EAP included tourism as one of its key sectors (Hashimoto, 1999), whilst many of the wider changes in environmental management that EU directives have set in place have helped to create situations from which recreation and tourism derive a clear benefit. This is essentially due to the close relationship between the nature and quality of the environment and its appeal to people at leisure. So, for example, EU concerns over marine pollution and the proper management of waste and effluent have helped to shape programmes for producing cleaner bathing waters and beaches around Europe's coastline.

European policy has also become more attuned to the negative impacts that production systems – especially in agriculture – may exert upon landscapes and wildlife. The Common Agricultural Policy, prior to its reform in the early 1990s, had been responsible for a range of changes in farming practice that had progressively destroyed habitats and devalued the aesthetic appeal of many rural landscapes: through intensification, specialisation and homogenisation of agrarian practice (Reynolds, 1998). Since then, however, a range of agri-environmental schemes have been introduced in an attempt to redress some of the imbalance between production and other interests. These schemes include the EU Set-Aside programme in which farmers have been paid to take some farmland out of production and put it to other uses (including recreation). Additionally, European governments have generally supported a range of smaller schemes aimed at fostering more sensitive management regimes in fragile environments. In Britain, Countryside Stewardship has been a comparatively successful example of a new initiative. Introduced in 1991, it allowed for a system of payments to farmers who were willing to:

- undertake conservation and/or restoration of important landscapes such as limestone grasslands, lowland heaths or wetlands, and salt marshes;
- develop access to restored sites for public enjoyment.

So although the primary objective of the scheme was environmental enhancement, recreation and tourism benefited from the linked requirement to improve access (Countryside Agency, 1998).

Similar policy relationships may be seen in European initiatives in regional development. As an essentially economic alliance, the EU has a strong record in regional policy and although economic motives provide the driving force, recreation and, especially, tourism have often been areas of consideration within European Regional Development Fund and – to a lesser extent – European Social Fund programmes. This has been evident in, for example, the RECHAR and LEADER programmes. RECHAR (which operated in two phases between 1990 and 1997) was aimed at the regeneration of former coalfield communities and allowed for the promotion of tourism as one of several areas that were available for project grant aid (Morgan and Jenkins, 1997). The development of industrial heritage – with which most RECHAR areas were well endowed – was a common theme, although in many cases expectations were not always matched by achievements (see, for example, Dicks, 1997). In contrast, the LEADER programmes were designed to provide alternative sources of income in areas of declining agricultural production and, unsurprisingly, development of rural recreation and tourism has been a prominent theme in many LEADER programme areas (see Hall and Page, 1999: 198–212 for a useful case study of a range of EU programmes – including LEADER – in Ireland).

Land designation

Land designation – whether for purposes of conservation and the protection of key resources, or as a means of providing for particular activities – is a policy approach that is now widely encountered across the globe. National parks, for instance, which constitute a prime example of land designation for conservational reasons, are to be found on all of the continents except for Antarctica (see Butler and Boyd, 2000). Although there is considerable variation in policy and practice (as a reflection of the natural fluctuations in national, regional or local patterns of demand, priority and opportunity), two basic approaches to designating land for recreation and tourism commonly occur:

1   The designation of tracts of land or specific facilities that are devoted to recreational and tourist use as primary functions and which enjoy general public access for purposes of leisure.
2   The provision of space and/or facilities that are set within resource areas that have a different primary function but which are able to accommodate recreation or tourism within smaller, dedicated areas that are set aside for such use.

In the UK, for example, the emphasis has tended to be placed upon the second of these approaches with most provision – especially in the countryside – occurring on land or water that is expected to fulfil a range of functions. Since 1945, the designation of land into a widening range of categories has become a primary mechanism through which government has sought to influence management practices on privately held, rural land. Typical forms of designation include the national parks, the AONBs, heritage coasts, national nature reserves (NNRs) and country parks, and when all forms of land designation are taken into account, almost half the rural area of England and Wales lies in one or other of the designated categories (Figure 6.2) (Sharpley, 1996).

However, owing to their plurality of purposes or, in some situations, the significance of areas such as NNRs as primary zones of conservation, the recreational and tourist provision in many of these designated zones has often been slight. Amongst the areas identified in Figure 6.2, only the country parks – which date from 1968 – constitute sites that are specifically declared for recreation or tourism. Recreational and tourist facilities clearly occur in the other designated categories – such as the national parks – and have become more numerous through time as levels of recreational activity have risen, but importantly, in no instance is leisure use the primary rationale of designation in these other categories.

The comparative shortage of land in the UK that is set aside for recreation and tourism use ensures, therefore, an emphasis upon provision in association with other activities. Here, changes in policy approaches have seen a significant growth in tourism and recreation sites in association with other land sectors. For example, following the Forestry Act (1972), there has been quite extensive development of provision within the forest holdings of the Forestry Commission (now the Forest Enterprise). This was especially evident in the period up to around 1990, by which time over 600 picnic sites, over 500 forest walks, more than 150 forest trails for riding and cycling and 21 visitor centres had been established (Glyptis, 1991b). The Forest Enterprise also owns a modest stock of self-catering holiday cottages on four sites in north-west Scotland (two sites), the North York Moors and Cornwall and manages some 22 caravan and camping sites for use by tourists, mostly within the New Forest and the Forest of Dean. In addition it is now an important provider of sites for specialist recreational activities such as motor sport and orienteering.

The modest level of provision of dedicated leisure sites in the UK is a product of a range of factors, including land-use history, patterns of landownership, a relative shortage of space and prevailing governmental policies that have tended to favour other sectors or concerns. In the USA, in contrast, significantly greater levels of land availability, high incidence of federal and state ownership of land and the presence of policy that has been disposed towards recreation and tourism, have each contributed to much higher levels of dedicated provision. The US example, therefore, is more in keeping with the first approach outlined above.

According to Siehl (2000) the US government in the 1950s was particularly prescient in recognising and responding to the needs of public outdoor recreation and providing funding (via the Land and Water Conservation Fund) for land acquisition and its management for leisure. Even though emerging concerns for environmental issues from around 1980 have subsequently

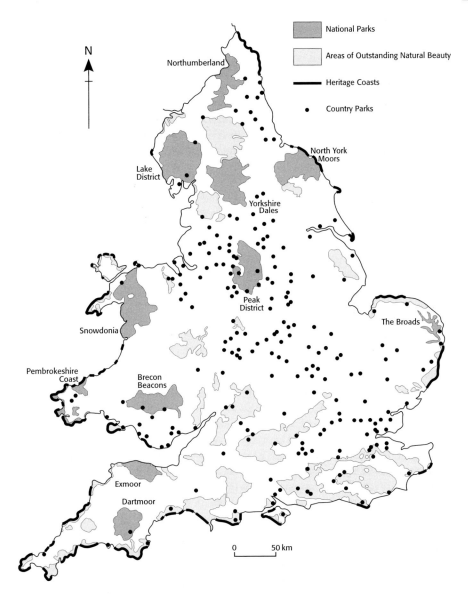

Figure 6.2 Designated land in England and Wales

undermined some of the strengths of US recreational policy, the achievements of federal, state and local authorities remain highly impressive.

From the 1960s onwards, the federal government augmented the existing system of national parks (which originated as long ago as 1872 with the designation of Yellowstone National Park in Wyoming) with an expanding area of national wildernesses, national trails, national recreation areas and national wild and scenic rivers. These continue to expand in area or extent. Since 1987, the national wilderness has increased by 15 per cent, the recreation areas by 11 per cent, the recreational trails by 14 per cent and the wild and scenic rivers by 41 per cent (Cordell and Betz, 2000). Moreover, as a major contrast to patterns in the UK, nearly all federally owned land and water is open for public recreational use although, with a heavy concentration of holdings in the more sparsely populated western USA, accessibility of some of the areas of recreational land and water is not always particularly high. Entry to sites is also generally subject to payment, which is a second point of contrast with UK practice.

Provision of dedicated land is also evident at the state level. State parks, forests, recreation areas, scenic drives and historic areas provide important components in the overall supply of recreational opportunity, often with a better spatial match to the distribution of demand than is evident in the case of federal holdings. As with federal lands, state holdings are substantial in extent – state forests, for example, covering more than 20 million ha and state parks some 3.2 million ha – and are continuing to expand. The total area of state-held recreational land in the USA grew by 32 per cent over the course of the 1990s (Cordell and Betz, 2000).

Integrated programmes of change

In a review essay on the theme of recreation and tourism planning in protected areas, McCool and Patterson (2000) identify several key trends. They note, for example, the increasing complexity of the patterns of linkage between people and resources, as the range of demands becomes more developed through time. They also draw attention to the ways in which recreation and tourism are being incorporated into a wider social agenda that relates to the enhancement of the quality of life. It is developments such as these that are promoting policies that take a more integrated approach in which provision for tourism and recreation become central, rather than ancillary, elements within more comprehensive programmes of change and where the recreational or tourist activity often contributes directly to the economic well-being of the host areas or enterprises.

There is an expanding range of examples that illustrate this process. In the urban environment, for example, the regeneration of former industrial locales has often leant heavily upon recreation and, especially, tourism as a force to drive economic development and the creation of new civic images (see below). At the same time in the countryside, recreation and tourism have become integral elements in the operation of a growing number of farms, whilst other resource areas – such as water reservoirs – have also seen their patterns of usage change. The development of recreation and tourism as a popular means of diversification of farming enterprise has already been outlined in Chapter 5. Across much of Europe, overproduction has encouraged a higher level of regulation of farming activity, and in the quest for supplementary (and complementary) sources of income, provision for recreation and tourism has become commonplace – especially in regions where farming is of a more marginal nature (see Chapter 5).

The wider integration of recreation and tourism into agriculture is primarily a reflection of changed conditions on the supply side. In contrast, integration of recreation and tourism into the management of water resources is essentially a response to changes in demand. In countries such as the UK, water-based leisure activity emerged as a major area of growth at the end of the 1960s and prompted the government of the time to use part of the Water Act of 1973 to place new requirements on the water authorities to provide for public recreation on reservoirs, wherever this could be achieved without risk to water supplies. The response has led to a progressive development of recreational provision as integral features in the management of lakes and reservoirs, especially in England and Wales. By the late 1970s, almost 90 per cent of over 530 reservoirs in England and Wales accommodated recreation on either the water or the shoreline (Tanner, 1977). Figure 6.3 illustrates the pattern of provision of major sites in Wales, where 118 out of a total of 137 lakes and reservoirs are now available to recreational and tourist visitors, amounting to over 12,000 ha of water space (Ballinger, 1996). Additionally, Figure 6.3 reveals the extent to which river systems have developed recreational uses that complement and extend the capacities of the lakes and reservoirs. In many instances, this also reflects a closer integration of recreational interests into the routine management of rivers.

Development of management techniques

Alongside the development of policy frameworks that are now more reflective of recreational and tourist demands, the ability to integrate recreation and tourism provision with other roles or functions of land or water space through the application of management techniques has been an important area of change. In particular, management has been obliged to find ways of resolving potential conflicts (whether between different areas of tourism and recreation, or with other activities or land uses), as well as addressing problems of detrimental impact on resources.

Several common approaches have been developed and become widely adopted in managing leisure sites. These include: the use of principles of compatibility; spatial and temporal zoning; and the designation of high-capacity 'honeypot' sites to help concentrate and segregate recreational or tourist visitors. Since these are generally well known, only an outline description is provided here.

Principles of compatibility

Central to much of the contemporary management of recreational or tourist conflict is an understanding of compatibility. Essentially this focuses upon recognition of categories of activity that are capable of coexisting in the same times and places without triggering conflicts, together with an identification of

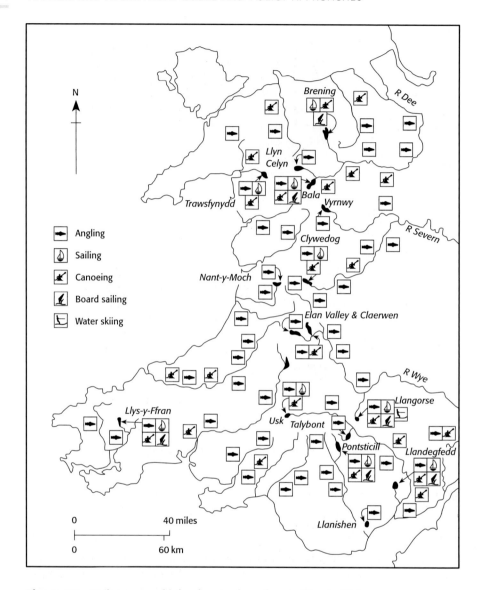

Figure 6.3 Major areas of inland water-based recreation in Wales
Source: Adapted from Ballinger (1996).

those activities where conflict is likely. This might be produced by problems such as visual intrusion of one event upon another, disturbance through noise, contrasting motivations of different user groups, or simple mismatches between the spatial requirements of competing activities. Thus, for example, contemplative recreations such as fishing will rarely coexist with active sports such as powerboating.

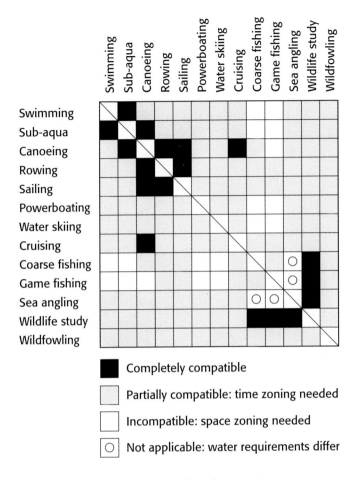

Figure 6.4 Compatibility matrix for water-based recreation
Source: After Patmore (1983).

Identification of likely sources of difficulty rests on a combination of management experience and intuition, but for organisational purposes, the construction of a simple compatibility matrix usually helps to identify groups of potential activities that are either capable of integration or which require segregation. Figure 6.4 provides an example of such a matrix, designed to help manage competing uses of water space by water-based recreations.

Spatial and temporal zoning

As Figure 6.4 implies, the application of an understanding of compatibility in practical terms generally centres upon the use of spatial or temporal zoning. This approach attempts to overcome problems of conflict by separating activities (or clusters of activity), either in space or in time.

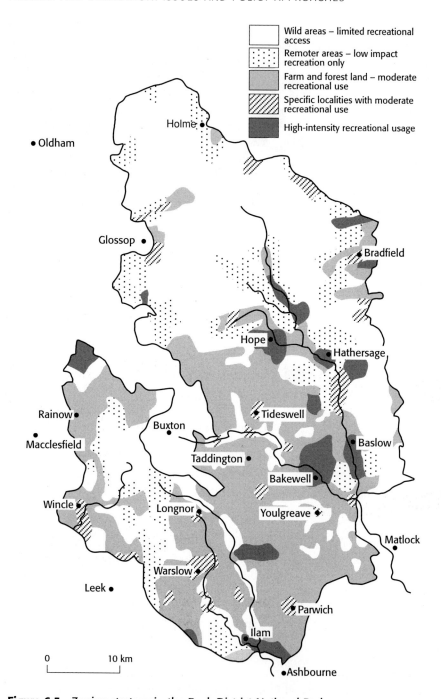

Figure 6.5 Zoning strategy in the Peak District National Park

RECREATION AND TOURISM: RESPONSES TO POLICY ISSUES   177

Spatial zoning entails the allocation of differing areas of a resource – for example, an area of water space or woodland – to different activities so that the physical separation reduces or removes the scope for one activity to interfere directly with another. It may also be applied when managers wish to protect particular areas from recreational usage that might exert unwanted impacts, for example, where there are overriding concerns for the conservation of a habitat. As a management method, therefore, spatial zoning may be applied at a range of spatial scales. For example, Figure 6.5 illustrates a spatial zoning policy applied to a national park area (the Peak District in northern England) that is intended to concentrate recreational and tourist activity into managed zones, whilst discouraging usage of more fragile, natural areas (Williams, 1998). In contrast, Figure 6.6 provides an example of spatial zoning at a much

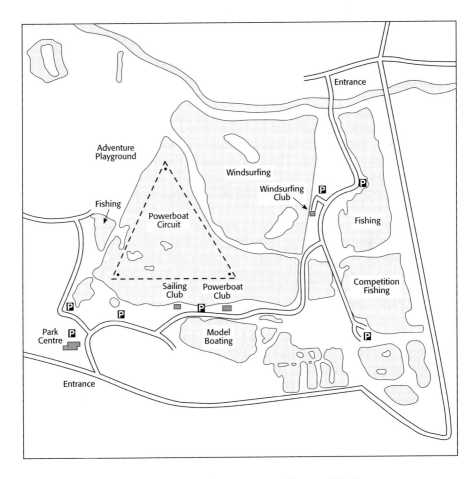

Figure 6.6 Zoning pattern at Kingsbury Water Park, Warwickshire
Source: After Ballinger (1996).

These issues arise, of course, as a consequence of the broad shift from an industrial to a post-industrial basis to urbanism across the developed world and which have exerted particularly significant impacts upon older, inner regions of industrial cities. Law (2000: 117) describes how 'increased competition arising from globalisation, combined with changing physical requirements for factories and more flexible forms of transport, affected many old industries and firms located on the edge of the city centre and in the surrounding inner-city areas, resulting in decline and closure'. Between 1971 and 1991 nearly 3 million urban manufacturing jobs were lost in Britain, contributing to an urban problem that was characterised by persistent unemployment, acute social deprivation and urban unrest (Hubbard, 1996).

The role of tourism and recreation as catalyst of regeneration has, however, emerged only recently. Law (1992) suggests that the use of tourism (in particular) in urban regeneration programmes diffused to Britain and Europe from North America. Tourism, he argues, was widely perceived as a growth industry that created significant levels of demand for labour and which therefore addressed a primary area of policy concern, especially during the mid-1980s. Tourism was seen by many city authorities as an appropriate way to speed up processes of recovery from deindustrialisation (van der Borg, 1992) and was capable of endowing local communities with a range of potential benefits. These included:

- social enhancement through the improvement to local amenities, cultural facilities, public transport and the fostering of civic pride;
- economic enhancement through job and wealth creation, attraction of inward investment and governmental redevelopment funds;
- environmental enhancement through new emphases upon conservation of the built environment, land reclamation programmes, removal of dereliction and the refashioning of urban space into new areas of amenity, such as cultural quarters (Evans, 2000).

The transformation of former industrial cities into tourist attractions required, however, significant enhancement in the image of the city and conscious investment in attractions and facilities. Short *et al.* (1993) emphasise the importance of reimaging in order to translate the often negative connotations of industrialism (past, old, work, pollution, production) into the positive associations of the post-industrial (future, new, leisure, unpolluted, consumption). Without such a transformation – without the revaluation of urban space (Meethan, 2001) – neither the necessary investment nor the tourists are likely to materialise.

A key feature of recent programmes of regeneration has been the use of public–private partnerships within integrated programmes of development that embrace tourism (Long, 1999). This is a reflection, first, of the complex realities of urban regeneration – in which many different elements need to be brought together and the interests of a range of stakeholders met. These include residents, tourists, local business operators, politicians and environmental or

RECREATION AND TOURISM: POLICY IN PRACTICE   181

conservation groups (Laws and le Pelley, 2000). Additionally, second, it points to important changes in approaches to urban governance. Several authors (for example, Hubbard, 1996; Meethan, 2001) draw attention to moves towards a reduced emphasis upon welfare approaches and the adoption of a more entrepreneurial stance that stresses new patterns of investment as a pathway to wealth creation. In this process a large number of partnerships were formed (for example, in London Dockland; Trafford, Manchester; the Don Valley, Sheffield; and the Greenwich waterfront) in which public enterprise often shaped an overall strategy and provided investment in basic infrastructure, whilst private capital frequently developed specific facilities and attractions (Law, 1992; Long, 1999).

Within many of the schemes that involved tourism, a common approach has been to develop clusters of attractions in what become well-defined concentrations of recreational or tourist activity. Some may be based in (or on the fringes) of city centres – for example, the Centenary Square development in Birmingham illustrated in Figures 4.1, 4.2 and 6.8 – others occur in zones of renewal, such as Port Solent or Swansea South Dock – illustrated in Figures 2.11 and 2.12. In general the pattern of provision combines prestige attractions (such as the National Indoor Arena in Birmingham) with supporting

Figure 6.8 The redevelopment of Birmingham's Gas Street Basin has seen new provision for canal cruising as a recreational and tourist activity as well as the development of canalside bars and restaurants. Modern hotel development (with the high-rise Jury's hotel to the left in this view and a Hyatt hotel out of shot to the right) has also been a key component in the redevelopment

infrastructure that centres upon leisure retailing, bars and restaurants, entertainment, cultural or heritage attractions and accommodation.

It is important to acknowledge, therefore, that whilst it is often tourism that is emphasised within leisure-based regeneration schemes, there are fundamental connections with local recreation. In many cases the development of tourism as part of a regenerative programme builds upon an existing stock of local facilities (e.g. public open space, cultural facilities, accommodation, retailing and food and drink services), but critically, it adds additional layers of infrastructure that service local recreational, as well as visiting tourist, needs. Thus, for example, Couch and Farr (2000) show how the visitors to Liverpool's museums and galleries – which include the prestigious Albert Dock redevelopment – comprise 65 per cent local residents, 20 per cent day visitors from outside and 15 per cent tourists who are staying in the city.

The following case study provides an example of an inner-area regeneration programme from the neighbouring city of Manchester, in which the development of recreational and tourist provision has been particularly prominent.

Case study: Urban regeneration – Manchester city centre

Manchester exhibits many of the characteristics of a former industrial city in transition. Deindustrialisation in the 1960s and 1970s created major zones of disused land and premises, whilst the restructuring of production from a manufacturing to a service economy led to significant problems with unemployment and associated social deprivation, particularly within the inner city. Recovery has been slow but, according to Law (2000: 122), 'in the late 1990s the city centre experienced a growing renaissance based upon leisure and the development of residences'.

Initial steps to creating the idea of Manchester as a tourist city were taken by the now defunct Greater Manchester Council in the early 1980s. Former railway sites were acquired at the disused Central Station and the Liverpool Road goods depot which, by 1986, had been redeveloped as a new Museum of Science and Industry (on part of the Liverpool Road site) and the G-Mex exhibition centre (at Central Station) (see Figure 6.9). This formed the basis for the subsequent development of the Castlefield district as a major zone of urban leisure and tourism, with the addition, after 1988, of:

- a new concert hall (the Bridgewater Hall);
- the Great Northern Warehouse (which includes a convention centre, restaurants, bars and shops);
- the Granada TV studio tours complex (an urban theme park featuring the set for the popular TV soap opera *Coronation Street*).

In 1993 the city launched a bid to host the 2000 Olympic Games that, although unsuccessful, attracted government grants to finance the building of the Indoor Arena on the northern side of the city centre. This has formed a focus

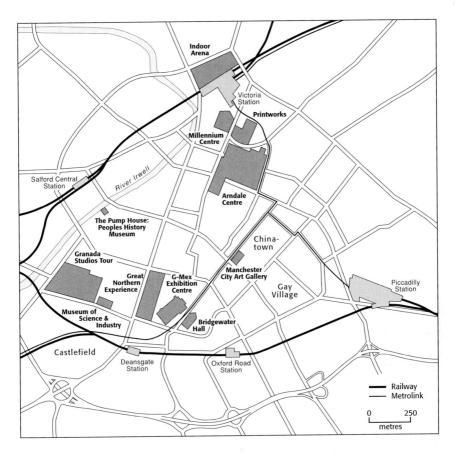

Figure 6.9 Visitor attractions and leisure areas in Manchester city centre
Source: After Law (2000).

for a secondary development of leisure facilities that includes the Printworks complex of cinemas, bars and restaurants and the Arndale shopping mall that was reconstructed following the IRA bomb attack of 1996.

Much of this development has been achieved through public–private partnerships, but significant initiatives have also emerged from the private sector alone. These include the creation of a major nightclub scene that has acquired a national reputation; the development of China Town as an important zone of restaurant operations with a strong cultural identity; and the creation of the Gay Village that has developed in a zone of former warehousing around the Canal Street area and which is reputed to attract visitors from a wide geographical area to its shops, bars and clubs.

Law (2000: 124) claims that, for the first time, the regeneration of Manchester's central area has been essentially led by leisure developments that have helped to transform the ambience of a decaying urban area and enhance its attraction to permanent residents. The permanent population has risen from a few hundred in the 1970s to over 10,000 by 2001, whilst the number and capacity of the central area hotels have grown from just 7 hotels (with 982 rooms) in 1980, to 24 hotels (with 3,231 rooms) in 2000. Tourist numbers, although fluctuating, are significant. In 1998, over 3.2 million people are thought to have visited Manchester for a total of nearly 11 million nights, spending over £400 million.

There are, however, evident problems with regeneration programmes based on leisure that the case of Manchester helps to illustrate. The higher-spending, international tourists have – perhaps predictably – proven to be hard to attract to a provincial city with a reputation based upon industrialism and an associated image problem. Maintaining a competitive position clearly requires continued improvement and enlargement of the visitor attractions base. Some attractions have also revealed only a limited capacity to stimulate repeat visiting so that, for example, initial levels of visiting to Granada Studios that touched 750,000 per annum soon fell to around 400,000. But as a base for regional forms of tourism and local patterns of recreation, it is evident that revitalised city centres such as Manchester are important arenas of development.

Source: Based on information from Law (2000).

Several general themes are illustrated by the example of regeneration. First, it reveals the increased recognition of recreation and tourism within certain areas of urban policy, reflecting the fact that tourism is now a significant economic force and that recreation is a primary feature of many people's daily lives. Recognition is mirrored, second, in a willingness to dedicate space and other resources to providing for recreation and tourism, particularly where wider benefits for the community at large are anticipated. Third, due to the complexity of the challenge that is often faced, urban regeneration generally reveals the importance of integrated approaches – especially through the reliance upon partnerships – and the willingness to tackle a range of interrelated issues to do with restoration and redevelopment of urban land or water space in a holistic fashion. Finally, many regeneration projects show how recreation and tourism may be managed to produce positive impacts across the physical, economic, social and environmental range. These are especially evident when redevelopment is concentrated into well-defined zones – as illustrated in the case of Manchester. These zones serve not only to minimise disruption to local residents by providing some measure of segregation of tourists, but more positively perhaps, help to create strong images and identities that can form powerful assets in urban place promotion.

Access to the countryside for recreation and tourism

The relative success of urban regeneration schemes that have featured recreational or tourist developments stems primarily from the fact that such schemes generally comprise initiatives that are spatially focused and which are shaped around measurable goals and outcomes. They also possess a strong economic dimension which tends to strengthen the commitment of policy-makers to supporting leisure provision as part of urban regeneration.

In contrast, the question of access to the countryside for recreational or tourist users poses a different and more complex challenge, and one where policy has been demonstrably less successful in addressing the issues that are raised. The complexity arises partly through the diffuse spatial patterns of demand for access that creates the need for a policy response that is spatially extensive. This has implications for resourcing. But additionally, the experience of managing countryside access in the UK illustrates several further issues, in particular:

- the impact of constraints that arise through patterns of landownership and which have tended to channel the development of access in particular directions;
- the consequences of the failure of policy-makers to dedicate adequate resources to providing access, so that access schemes that have been set in place often achieve only partial success in meeting their objectives;
- an inability to develop a comprehensive approach that actually extends access to the population at large.

The situation in the UK presents a number of contrasts with systems that are seen elsewhere. In North America, as was discussed earlier, extensive patterns of access have been provided within dedicated federal and state recreational areas, as an integral component in the development and management of these zones. Alternatively, in Norway, Sweden, Denmark and Germany, for example, traditional rights of use have been translated into a pattern whereby the public has a general right of access to rural land for recreation, within limitations that are posed by a need to protect nature, landscape or agricultural crops, or respect certain rights of property and privacy. In these countries the land is normally regarded as a national resource to be used for the benefit of society, as well as the owner (Scott, 1998). In the UK, the Scottish Parliament is currently legislating to provide for a system that is essentially similar to these European models – in the use of a presumption of a general right of access to land (Davidson, 2000). In the rest of England and Wales, however, rural access continues to be shaped around a complex, multilayered pattern of provision that owes as much to accidents of history as it does to contemporary need (Kay, 2002).

The current range of access arrangements in England and Wales includes, as established elements:

- a core provision of some 193,000 km of public rights of way, most of which are paths but which also include bridleways;
- the country parks and picnic sites that were created following the 1968 Countryside Act;
- free access to some 200,000 ha of open land in the ownership of the National Trust;

- access to some areas of land held by the Forest Authority and the water utilities;
- over 34,000 ha of open access land negotiated by local authorities with landowners under the terms of the 1949 National Parks and Access to the Countryside Act (Curry, 1994; Ravenscroft, 1999).

More recently, this pattern has been supplemented by newly negotiated access agreements within agri-environmental schemes such as Countryside Stewardship and, more significantly, the extension of public rights of access to some 1.6 million ha of open land (mostly in upland areas) under the terms of the Countryside and Rights of Way Act 2000.

Outwardly, this pattern of provision – although fragmented across a range of different forms of access – appears both comprehensive and generous, reflecting a developing commitment on the part of government to extending access as the demands of recreation and tourism have grown. In practice, however, there are significant deficiencies in both the extent of provision and the effectiveness with which it matches the pattern of people's needs. This, it is argued, is a consequence of the exercise of the vested powers of landowning interests, combined with an extended history of rather tentative policy-making. In this latter context, problems of legal regulation, a tendency to produce policy proposals that have proven to be too ambitious, a susceptibility to reflecting the interests of vocal minority groups, and shortcomings in the promotion of some of the newer forms of access, have all emerged as limitations.

The capacity of landowning political lobbies to shape policy in their favour has been noted earlier in this chapter. This tacit acceptance of the rights of landowners has become quite firmly entrenched in governmental approaches to rural access which, with the possible exception of the Countryside and Rights of Way Act (2000), has been characteristically deferential. Ravenscroft (1996) draws attention to the way in which landowners have tended to construct the question of access to private land as an encroachment upon their libertarian freedom. This has encouraged the adoption of a paternalistic perspective – one in which landowners who accede to pressures for public access to their land do so as a form of 'gift' and where access arrangements are entered into voluntarily and with due compensation from the public purse for any loss, whether to the productive capacity of land or in the more elusive concept of privacy. This traditional view of the public–private partnership in securing access to rural land continues to be influential (in schemes such as Countryside Stewardship) but critically, has tended to perpetuate a pattern in which access is always temporary and contingent upon negotiated agreement and public subsidy. Ravenscroft (1999) criticises this approach as allowing landowners to continue to view the question of access in purely technical terms, i.e. permitting access only in designated areas, under supervision (in the form of managed use) and for approved activities. In the process, the wider question of whether such approaches address social needs for rural recreation and tourism remains neglected.

In light of the limitations that ownership creates and the limited impact of initiatives such as country parks (Curry, 1994), the focus of recent access policy has tended to fall upon the rights of way network. In the 1949 National Parks Act, government attempted to strengthen this ancient system by creating new levels of responsibility for local authorities for the identification and recording of rights of way in definitive maps. Legal procedures for establishing and amending rights of way have also been developed through several Parliamentary Acts. However, in practice, regulation of the rights of way system has suffered from several related problems, including:

- a lack of resourcing, often – it has been argued – by a process of marginalisation of rights of way interests by appending them to highways legislation and organising the practical work within highways departments of local authorities (Curry, 1994);
- a problem of slow progress on the preparation of definitive maps, many of which remain incomplete at the time of writing and some likely to remain so well into the future;
- complexity in the legal processes for extending or amending rights of way through Public Path Orders and in the conduct of appeals to decisions related to these orders.

It has also become evident that the public at large are characteristically uncertain of their legal rights in respect of path usage and how rights of access may be asserted. Furthermore, where paths are poorly signed or where they may have become overgrown, waterlogged or ploughed, these conditions are often a powerful deterrent to use that will confound all but the most well-equipped and determined visitors.

In aiming to address problems of this nature, the Countryside Commission launched several major initiatives between 1987 and 1992 with the objective of enhancing public access through traditional rights of way. First, it proposed a new, hierarchical arrangement of paths into national trails, regional routes and local paths, with the latter subdivided into parish paths (which were not to be promoted) and local walks (which were to be promoted) (Countryside Commission, 1988). Linked to these proposals, second, were the contentious objectives that by the end of the century, the entire rights of way network would be legally defined, properly maintained and well publicised (Countryside Commission, 1987). In so doing, however, it set in place a policy that was hopelessly overambitious and, some have argued, inappropriate. Kay (1989) describes these proposals as unrealistic, unnecessary and unhelpful: unrealistic as there was no reliable means of resourcing the exercise; unnecessary in that preoccupation with the total network (much of which is virtually unused) distracts attention from a more focused pattern of investment in routes that have evident merit and use; unhelpful because it appeared to align the Commission with the interests of a small number of crusaders within bodies such as the Ramblers' Association who were committed to the total recovery of the rights of way network.

The lack of realism in the proposals soon became widely evident and the Commission attempted to produce a revised approach (whilst still maintaining the spirit of its original proposals) in its 'Milestones Approach' (Countryside Commission, 1993c). This aimed to instil a more strategic direction to the initiative in which local authorities would work towards the national targets within a year-by-year programme of projects that would aim for a progressive enhancement of the legal definition, condition and promotion of paths. However, the informal designation of this scheme by many rights of way officers as the 'millstones approach' provides some indication of the burdensome nature of the task and the lack of realism that has surrounded its delivery.

This links to a third issue – empowerment. Kay (2002) argues that much of the recent development in access through rights of way, whilst paying lip-service to notions of widening public access, has often produced policies that have an exclusionary effect. Kay asserts that this is partly due to the disproportionate attention accorded to minority pressure groups who command attention because they are conversant with access issues, possess knowledge and who have vested interests in making their views known, but who do not represent the needs and interests of the public at large. As a result, policy has tended to favour these minorities, rather than focusing upon ways of helping a wider public gain access to the countryside.

This tendency is perhaps most clearly evident in debates that have centred on the most recent access initiative in England and Wales, the creation of a right to roam over open land – a concession that forms the cornerstone of the Countryside and Rights of Way Act. Ravenscroft (1999) notes that the long-standing campaign by groups such as the Ramblers' Association for a right to roam is both an expression of ideology – a desire to restore what is perceived by some as a traditional right that was widely extinguished by processes of agricultural improvement and industrialisation – as well as a question of practicality for walking enthusiasts in gaining access to new and remote environments (Crofts, 1995). Kay (1998, 2002), however, draws a more critical line, arguing that the right to roam represents the interests of only a committed and experienced minority of enthusiasts. It risks extending problems of damage and erosion beyond the limited capacities of managers to deal with the problems, and diverts scarce resources into supporting a system that favours the few at the expense of the majority. As Kay and Moxham (1996) make clear, walking as a recreational or tourist activity embraces a wide range of purposes, motives and competencies and provision that favours only those who possess skills, knowledge and experience – as walking in open country inevitably requires – acts to exclude rather than extend effective public access.

Part of the proposals for access to open land under the Countryside and Rights of Way Act require the Countryside Agency to record, in maps, the extent of open access land. This is an important first step to raising public awareness of new opportunities, but it is one that has not always been taken. Ravenscroft (1999), in criticising certain aspects of some recent agri-environmental schemes such as Countryside Stewardship, argues that they produce a form of hyper-reality as the new provision remains unrecognised by a majority of citizens. Since many of these schemes exist in only a temporary form, there has been an official reluctance to record their presence on maps; most are unpromoted and, with a few exceptions, most provide no clear guidance as to how they may be used and for what purposes. Without effective promotion, however, present initiatives aimed at widening public access to the countryside amongst the millions of inexperienced, hesitant and only occasional visitors are likely to founder. Hence, as Kay (2002: 252) concludes, 'the politics of promotion therefore deserve no less attention than the politics of provision'.

Conclusion

This chapter has explored a range of issues that impinge upon policy developments that affect recreational and tourist opportunity. The discussion has shown how recreation and tourism have been adopted only partially and often belatedly into broader areas of policy concern. This has occurred for a number of reasons, but particular emphasis has been placed upon the tendency to view recreation and tourism as residual areas of policy interest and where, consequently, there has often been a marked reluctance to address issues that surround the allocation of resources or the hegemony of resource ownership, especially land. However, as the full implications of the transformations of post-industrial production and consumption have become evident, so a greater significance has been accorded to recreation and tourism as the scope for incorporation of leisure activity into areas such as urban and rural restructuring has been recognised. As providers have acquired an enhanced understanding of recreational or tourist requirements and the impacts that these activities create, so they have become better equipped to manage key resource areas to support a diverse range of recreational or tourist activities. These developments have produced a significant extension in the opportunity for public leisure over the last two decades or so, even though – as the example of access to the countryside illustrates – the best intentions of the policy-makers have sometimes fallen short of the desired goals.

In the context of the overall thesis that has been pursued in this book, the preceding discussion suggests that although recreation and tourism draw upon rather different traditions, the processes of de-differentiation that have been explored in earlier chapters mean that in policy areas too, distinctions between recreation and tourism are becoming progressively less relevant and less meaningful. In some contexts – such as within economic development or urban renewal – the economic power of tourism attracts special attention and targeted policies, but in many other situations – such as countryside access or the management of national parks – recognition of *difference* between recreational or tourist visitors is seldom an issue that shapes policy approaches. In so far as

Bar-On, R.R. (1997) 'Global tourism trends – to 1996', *Tourism Economics*, **3** (3): 289–300.

Beioley, S. (1999) 'Short and sweet – the UK short-break market', *Insights*, **10** (B): 63–78.

Bennington, J. and White, J. (eds) (1988) *The Future of Leisure Services*, Harlow: Longman.

Billinge, M. (1996) 'A time and a place for everything: an essay on recreation, reCreation and the Victorians', *Journal of Historical Geography*, **22** (4): 443–59.

Blank, U. (1996) 'Tourism in United States cities', in Law, C.M. (ed.), *Tourism in Major Cities*, London: International Thomson Business, pp. 206–32.

Blunden, J. and Curry, N. (1990) *A People's Charter: Forty years of the National Parks and Access to the Countryside Act*, London: HMSO.

Boniface, P. and Fowler, P.J. (1993) *Heritage and Tourism in the Global Village*, London: Routledge.

Boorstin, D. (1964) *The Image: A guide to pseudo-events in America*, New York: Harper.

Borsay, P. (1989) *The English Urban Renaissance: Culture and society in the provincial town, 1660–1770*, Oxford: Clarendon Press.

Bote Gomez, V. and Sinclair, M.T. (1996) 'Tourism demand and supply in Spain', in Barke, M., Towner, J. and Newton, M.T. (eds), *Tourism in Spain: Critical issues*, Wallingford: CAB International, pp. 65–88.

Bowler, I., Clark, G., Crockett, A., Ilbery, B. and Shaw, A. (1996) 'The development of alternative farm enterprises: a study of family labour farms in the north Pennines of England', *Journal of Rural Studies*, **12** (3): 285–95.

Bowler, I. and Strachan, A.J. (1976a) *Parks and Gardens in Leicester*, Leicester: Leicester City Council Recreational and Cultural Services Department.

Bowler, I. and Strachan, A.J. (1976b) 'Visitor behaviour in urban parks', *Parks and Recreation*, **41** (9): 18–25.

Bramham, P., Hendry, I., Mommaas, H. and van der Poel, H. (eds) (1989) *Leisure and Urban Processes. Critical Studies of Leisure Policy in Western European Cities*, London: Routledge.

Bray, R. (1996) 'The package holiday market in Europe', *Travel and Tourism Analyst*, **4**: 51–71.

Breedveld, K. (1996) 'Post-Fordist leisure and work', *Loisir et Société*, **19** (1): 67–90.

Brent Ritchie, J.R. and Goeldner, C. (eds) (1987) *Travel, Tourism and Hospitality Research: A handbook for managers and researchers*, Chichester: John Wiley.

Briassoulis, H. and van der Straaten, J. (eds) (1992) *Tourism and the Environment: Regional, economic and policy issues*, Dordrecht: Kluwer Academic.

British Travel Association (1969) *Patterns in British Holiday-making 1951–1968*, London: British Travel Association.

Bromley, R.D.F. and Thomas, C. (eds) (1992) *Retail Change: Contemporary issues*, London: UCL Press.

BTA (British Tourist Authority) (1995) *Digest of Tourist Statistics No. 18*, London: British Tourist Authority.

BTA (2001) *Digest of Tourist Statistics No. 24*, London: British Tourist Authority.

Buhalis, D. (1998) 'Strategic use of information technologies in the tourism industry', *Tourism Management*, **19** (5): 409–21.

Bull, P. and Church, A. (1996) 'The London tourism complex', in Law, C.M. (ed.), *Tourism in Major Cities*, London: International Thomson Business, pp. 155–79.

Bunce, M. (1994) *The Countryside Ideal: Anglo-American images of landscape*, London: Routledge.
Burgers, J. (1995) 'Public space in the post-industrial city', in Ashworth, G.J. and Dietvorst, A.G.J. (eds), *Tourism and Spatial Transformations: Implications for Policy and Planning*, Wallingford: CAB International.
Burgess, J. (1995) *The Politics of Trust: Reducing fear of crime in urban parks*, Park Life Working Paper No. 8, Stroud: Comedia Demos.
Burgess, J., Harrison, C.M. and Limb, M. (1988) 'People, parks and the urban green: a study of popular meanings and values for open spaces in the city', *Urban Studies*, 25 (6): 455–73.
Burtenshaw, D., Bateman, M. and Ashworth, G.J. (1981) *The City in Western Europe*, Chichester: John Wiley.
Burtenshaw, D., Bateman, M. and Ashworth, G.J. (1991) *The City in Western Europe*, London: David Fulton.
Burton, R.C.J. (1994) 'Geographical patterns of tourism in Europe', in Cooper, C.P. and Lockwood, A. (eds), *Progress in Tourism Recreation and Hospitality Management*, vol. 5, Chichester: John Wiley, pp. 3–25.
Busby, G. and Rendle, S. (2000) 'The transition from tourism on farms to farm tourism', *Tourism Management*, 21 (6): 635–42.
Butler, R.W. (1991) 'West Edmonton Mall as a tourist attraction', *The Canadian Geographer*, 35 (3): 287–95.
Butler, R.W. and Boyd, S.W. (eds) (2000) *Tourism and National Parks: Issues and implications*, Chichester: John Wiley.
Butler, R., Hall, C.M. and Jenkins, J.M. (eds) (1999) *Tourism and Recreation in Rural Areas*, Chichester: John Wiley.
Central Statistical Office (1990) *Social Trends 20*, London: HMSO.
Chadwick, G.F. (1966) *The Park and the Town*, London: The Architectural Press.
Chadwick, R. (1987) 'Concepts, definitions and measures used in travel and tourism research', in Brent Ritchie, J.R. and Goeldner, C. (eds), *Travel, Tourism and Hospitality Research: A handbook for managers and researchers*, Chichester: John Wiley, pp. 47–61.
Cherry, G.E. (1984) 'Leisure and the home: a review of changing relationships', *Leisure Studies*, 3 (1): 35–52.
Cherry, G.E. and Rogers, A. (1996) *Rural Change and Planning: England and Wales in the twentieth century*, London: E. & F.N. Spon.
Clammer, J. (1992) 'Aesthetics of the self: shopping and social being in contemporary Japan', in Shields, R. (ed.), *Lifestyle Shopping: The subject of consumption*, London: Routledge, pp. 195–215.
Clark, G., Darrall, J., Grove-White, R., Macnaghten, P. and Urry, J. (1994) *Leisure Landscapes. Leisure, Culture and the English Countryside: Challenges and conflicts*, Lancaster University: Centre for the Study of Environmental Change.
Clarke, J. and Crichter, C. (1985) *The Devil Makes Work: Leisure in capitalist Britain*, Basingstoke: Macmillan.
Clawson, M. (1963) *Land and Water for Recreation: Opportunities, problems and policies*, Chicago: Rand McNally.
Cloke, P. (1993) 'The countryside as commodity: new rural spaces for leisure', in Glyptis, S. (ed.), *Leisure and the Environment: Essays in honour of Professor J.A. Patmore*, London, Belhaven, pp. 53–67.
Cloke, P. and Little, J. (1996) *Contested Countryside Cultures*, London: Routledge.

Cockerell, N. (1997) 'Urban tourism in Europe', *Travel and Tourism Analyst*, 6: 44–67.
Cohen, E. (1972) 'Towards a sociology of international tourism', *Social Research*, 39 (1): 64–82.
Cohen, E. (1974) 'Who is a tourist? A conceptual clarification', *Sociological Review*, 22 (4): 527–55.
Conway, H. (1991) *People's Parks: The design and development of Victorian parks in Britain*, Cambridge: Cambridge University Press.
Cooper, C.P. (ed.) (1989) *Progress in Tourism, Recreation and Hospitality Management*, vol. 1, London: Belhaven.
Cooper, C.P. (ed.) (1990) *Progress in Tourism, Recreation and Hospitality Management*, vol. 2, London: Belhaven.
Cooper, C.P. (ed.) (1991) *Progress in Tourism, Recreation and Hospitality Management*, vol. 3, London: Belhaven.
Cooper, C.P. (1997) 'Parameters and indicators of decline of the British seaside resort', in Shaw, G. and Williams, A.M. (eds), *The Rise and Fall of British Coastal Resorts*, London: Mansell, pp. 79–101.
Cooper, C.P. and Lockwood, A. (eds) (1994) *Progress in Tourism Recreation and Hospitality Management*, vol. 5, Chichester: John Wiley.
Coppock, J.T. and Duffield, B. (1975) *Recreation in the Countryside*, London: Macmillan.
Corbin, A. (1995) *The Lure of the Sea*, London: Penguin.
Cordell, H.K. and Betz, C.J. (2000) 'Trends in outdoor recreation supply on public and private lands in the US', in Gartner, W.C. and Lime, D.W. (eds), *Trends in Outdoor Recreation, Leisure and Tourism*, Wallingford: CAB International, pp. 75–89.
Couch, C. and Farr, S. (2000) 'Museums, galleries, tourism and regeneration: some experiences from Liverpool', *Built Environment*, 26 (2): 152–63.
Countryside Agency (1998) *Countryside Stewardship: Monitoring and Evaluation of the Pilot Scheme*, Countryside Agency Research Note No. 3, Cheltenham: Countryside Agency.
Countryside Agency (1999) *The State of the Countryside*, Cheltenham: Countryside Agency.
Countryside Commission (1985) *National Countryside Recreation Survey 1984*, Cheltenham: Countryside Commission.
Countryside Commission (1987) *Recreation 2000: Enjoying the countryside*, Cheltenham: Countryside Commission.
Countryside Commission (1988) *Paths, Routes and Trails: A consultation paper*, Cheltenham: Countryside Commission.
Countryside Commission (1993a) *The National Park Authority: Purposes, powers and administration*, Cheltenham: Countryside Commission.
Countryside Commission (1993b) *Principles for Tourism in the Countryside*, Cheltenham: Countryside Commission.
Countryside Commission (1993c) *National Targets for Rights of Way: The milestones approach*, Cheltenham: Countryside Commission.
Countryside Commission (1995) *National Survey of Countryside Recreation 1990: Summary of results*, Cheltenham: Countryside Commission.
Countryside Commission (1996) *Visitors to National Parks: Summary of the 1994 Survey Findings*, Cheltenham: Countryside Commission.
Craven, E. (2000) 'Bluewater: retail tourism in the South East', *Insights*, 11 (C): 37–46.

CRN (Countryside Recreation Network) (1996) *UK Day Visits Survey 1994*, Cardiff: Countryside Recreation Network.
Crofts, R. (1995) 'Natural heritage zones: a new approach in Scotland', in Fladmark, J.M. (ed.), *Sharing the Earth: Local identity in global culture*, London: Donhead, pp. 227–44.
Crouch, D. (1995) *The Popular Culture of City Parks*, Park Life Working Paper No. 9, Stroud: Comedia Demos.
Crouch, D. (ed.) (1999) *Leisure/Tourism Geographies*, London: Routledge.
Cunningham, H. (1980) *Leisure in the Industrial Revolution c1780–c1880*, London: Croom Helm.
Curry, N. (1994) *Countryside Recreation, Access and Land Use Planning*, London: E. & F.N. Spon.
Dartmoor National Park Authority (1991) *Dartmoor National Park Plan 2nd Review*, Bovey Tracey: Dartmoor National Park Authority.
Dartmoor National Park Authority (1997) *Annual Report 1996–7*, Newton Abbot: Dartmoor National Park Authority.
Davidson, R. (1992) *Tourism in Europe*, London: Pitman.
Davidson, R. (2000) 'Proposals for access legislation in Scotland', *Countryside Recreation*, **8** (2): 7–11.
Davidson, R. and Maitland, R. (1997) *Tourism Destinations*, London: Hodder and Stoughton.
Deem, R. (1986) *All Work and No Play? The Sociology of Women and Leisure*, Milton Keynes: Open University Press.
Defoe, D. (1971) *A Tour through the Whole Island of Great Britain*, Harmondsworth: Penguin.
Demars, S. (1990) 'Romanticism and American national parks', *Journal of Cultural Geography*, **11** (1): 17–24.
Denman, R. (1994) 'The farm tourism market', *Insights*, **6** (B): 49–64.
Dicks, B. (1997) 'Regeneration versus representation in the Rhondda: the story of the Rhondda Heritage Park', *Contemporary Wales*, **9**: 56–73.
DoE (Department of Environment) (1990) *This Common Inheritance*, London: HMSO.
Durie, A.J. (1994) 'The development of the Scottish coastal resorts in the central lowlands circa 1770–1880. From Gulf Stream to Golf Stream', *The Local Historian*, **24** (4): 206–16.
Dwyer, J. and Hodge, I. (1996) *Countryside in Trust: Land management by conservation, recreation and amenity organisations*, Chichester: John Wiley.
Edmunds, M. (1999) 'Rural tourism in Europe', *Travel and Tourism Analyst*, **6**: 37–50.
Edwards, J.A. (1996) 'Waterfronts, tourism and economic sustainability: the United Kingdom experience', in Priestley, G.K. *et al.* (eds), *Sustainable Tourism? European Experiences*, Wallingford: CAB International, pp. 86–98.
Edwards, R. (1991) *Fit for the Future: Report of the National Parks Review Panel*, Cheltenham: Countryside Commission.
English Tourism Council (2000) 'Tourism by UK residents in 1997', *Insights*, **11** (F): 7–12.
Evans, G. (2000) 'Planning for urban tourism: a critique of borough development plans and tourism policy in London', *International Journal of Tourism Research*, **2** (5): 307–26.

Evans, N. and Ilbery, B.W. (1992) 'The distribution of farm-based accommodation in England and Wales', *Journal of the Royal Agricultural Society of England*, **153**: 67–80.
Falk, P. and Campbell, C. (eds) (1997) *The Shopping Experience*, London: Sage.
Featherstone, M. (1991) *Consumer Culture and Postmodernism*, London: Sage.
Fennell, D.A. and Weaver, D.B. (1997) 'Vacation farms and ecotourism in Saskatchewan, Canada', *Journal of Rural Studies*, **13** (4): 467–75.
Fladmark, J.M. (ed.) (1995) *Sharing the Earth: Local identity in global culture*, London: Donhead.
Fowler, P.J. (1992) *The Past in Contemporary Society*, London: Routledge.
Frater, J.M. (1983) 'Farm tourism in England: planning, funding, promotion and some lessons from Europe', *Tourism Management*, **4** (3): 167–79.
Gartner, W.C. and Lime, D.W. (eds) (2000) *Trends in Outdoor Recreation, Leisure and Tourism*, Wallingford: CAB International.
Gershuny, J. and Jones, S. (1987) 'The changing work/leisure balance in Britain, 1961–1984', in Horne, J. et al. (eds), *Sport, Leisure and Social Relations*, London: Routledge, pp. 9–50.
Gerstl, J. (1991) 'Routine and extraordinary leisure', *Loisir et Société*, **14** (2): 373–80.
Gilbert, D.C. (1990) 'Conceptual issues in the meaning of tourism', in Cooper, C.P. (ed.), *Progress in Tourism, Recreation and Hospitality Management*, vol. 2, London: Belhaven, pp. 4–27.
Glyptis, S.A. (1991a) 'Sport and tourism', in Cooper, C.P. (ed.), *Progress in Tourism, Recreation and Hospitality Management*, vol. 3, London: Belhaven, pp. 165–83.
Glyptis, S.A. (1991b) *Countryside Recreation*, Harlow: Longman.
Glyptis, S.A., McInnes, H. and Patmore, J.A. (1987) *Leisure in the Home*, London: Sports Council/ESRC.
Glyptis, S. (ed.) (1993) *Leisure and the Environment: Essays in honour of Professor J.A. Patmore*, London: Belhaven.
Go, F.M. (1992) 'The role of computerized reservation systems in the hospitality industry', *Tourism Management*, **13** (1): 22–6.
Gold, J.R. and Ward, S.V. (eds) (1994) *Place Promotion: The use of publicity and marketing to sell towns*, Chichester: John Wiley.
Goodall, B. (1992) 'Coastal resorts: development and re-development', *Built Environment*, **18** (1): 5–11.
Goodhead, T., Kasic, N. and Wheeler, C. (1996) 'Marinas and yachting', in Goodhead, T. and Johnson, D. (eds), *Coastal Recreation Management. The Sustainable Development of Maritime Leisure*, London: E. and F.N. Spon, pp. 181–200.
Goodhead, T. and Johnson, D. (eds) (1996) *Coastal Recreation Management. The Sustainable Development of Maritime Leisure*, London: E. and F.N. Spon.
Gordon, C. (1997) 'Oasis forest villages: developing with the environment', *Insights*, **9** (C): 19–28.
Gordon, C. (1998) 'Holiday centres: responding to the consumer', *Insights*, **9** (B): 1–11.
Goss, J. (1993) 'The "Magic of the Mall": an analysis of form, function and meaning in the contemporary retail built environment', *Annals of the Association of American Geographers*, **83** (1): 18–47.
Graburn, N.H.H. (1983) 'The anthropology of tourism', *Annals of Tourism Research*, **10** (1): 9–33.

Graham, B., Ashworth, G.J. and Tunbridge, J.E. (2000) *A Geography of Heritage: Power, culture and economy*, London: Arnold.
Green, N. (1990) *The Spectacle of Nature: Landscape and bourgeois culture in 19th century France*, Manchester: Manchester University Press.
Greenhalgh, L. and Worpole, K. (1995) *Park Life: Urban parks and social renewal*, Stroud: Comedia Demos.
Gunn, C.A. (1988) *Tourism Planning*, New York: Taylor and Francis.
Haggard, L.M. and Williams, D.R. (1992) 'Identity affirmation through leisure activities: leisure symbols of the self', *Journal of Leisure Research*, 24 (1): 1–18.
Hall, C.M. (1997) 'Mega-events and their legacies', in Murphy, P.E. (ed.), *Quality Management in Urban Tourism*, Chichester: John Wiley, pp. 75–87.
Hall, C.M. (2000) *Tourism Planning: Policies, processes and relationships*, Harlow: Addison-Wesley Longman.
Hall, C.M. and Page, S.J. (1999) *The Geography of Tourism and Recreation: Environment, place and space*, London: Routledge.
Harre, R. (1990) 'Leisure and its varieties', *Leisure Studies*, 9 (3): 187–96.
Harrison, C. (1991) *Countryside Recreation in a Changing Society*, London: TMS Partnership.
Harvey, D. (1989) *The Condition of Postmodernity*, Oxford: Blackwell.
Hashimoto, A. (1999) 'Comparative evolutionary trends in environmental policy: reflections on tourism development', *International Journal of Tourism Research*, 1 (3): 195–216.
Hass-Klau, C. (1990) *The Pedestrian and City Traffic*, London: Belhaven.
Haworth, J.T. (1986) 'Meaningful activity and psychological models of non-employment', *Leisure Studies*, 5 (3): 281–98.
Henderson, K.A. (1990) 'The meaning of leisure for women: an integrative review of the research, *Journal of Leisure Research*, 22 (3): 228–43.
Heung, V.C.S. and Qu, H. (1998) 'Tourism shopping and its contributions to Hong Kong', *Tourism Management*, 19 (4): 383–6.
Hiller, H.H. (2000) 'Mega-events, urban boosterism and growth strategies: an analysis of the objectives and legitimations of the Cape Town 2004 Olympic bid', *International Journal of Urban and Regional Research*, 24 (2): 439–58.
Hinch, T.D. and Higham, J.E.S. (2001) 'Sport tourism: a framework for research', *International Journal of Tourism Research*, 3 (1): 45–58.
Holt, R. (1985) 'The bicycle, the Bourgeoisie and the discovery of rural France, 1880–1914', *British Journal of Sports History*, 2 (2): 127–39.
Hopkins, J. (1991) 'West Edmonton Mall as a centre for social interaction', *The Canadian Geographer*, 35 (3): 268–79.
Hopkins, J. (1999) 'Commodifying the countryside: marketing myths of rurality', in Butler, R., Hall, C.M. and Jenkins, J.M. (eds), *Tourism and Recreation in Rural Areas*, Chichester: John Wiley, pp. 139–56.
Horne, J., Jary, D. and Tomlinson, A. (eds) (1987) *Sport, Leisure and Social Relations*, London: Routledge.
Horner, S. and Swarbrook, J. (1996) *Marketing Tourism, Hospitality and Leisure in Europe*, London: International Thomson Business.
Howkins, A. and Lowerson, J. (1979) *Trends in Leisure 1919–1939*, London: Sports Council/SSRC.
Hubbard, P. (1996) 'Re-imaging the city: the transformation of Birmingham's urban landscape', *Geography*, 81 (1): 26–36.

Hunter, C. and Green, H. (1995) *Tourism and the Environment: A sustainable relationship?*, London: Routledge.

Ilbery, B.W. (1989) 'Farm-based recreation: a possible solution to falling farm incomes?', *Journal of the Royal Agricultural Society of England*, 150: 57–67.

Ilbery, B.W. (1991) 'Farm diversification as an adjustment strategy on the urban fringe of the West Midlands', *Journal of Rural Studies*, 7 (3): 207–18.

Inskeep, E. (1991) *Tourism Planning: An integrated and sustainable development*, New York: Van Nostrand Reinhold.

Iso-Ahola, S.E. (1982) 'Towards a social psychological theory of tourism motivation', *Annals of Tourism Research*, 9 (2): 256–62.

Jackson, E.L. (1991) 'Shopping and leisure: implications of West Edmonton Mall for leisure and for leisure research', *The Canadian Geographer*, 35 (3): 280–7.

James, K. (2001) ' "I've just gotta have my own space!" The bedroom as a leisure site for adolescent girls', *Journal of Leisure Research*, 33 (1): 71–90.

Jansen, A.C.M. (1989) 'Funshopping as a geographical notion, or the attraction in inner-city Amsterdam as a shopping area', *Tijdschrift voor Economische en Sociale Geografie*, 80 (3): 171–83.

Jansen-Verbeke, M. (1986) 'Inner-city tourism: resources, tourists and promoters', *Annals of Tourism Research*, 13 (2): 79–100.

Jansen-Verbeke, M. (1991) 'Leisure shopping: a magic concept for the tourism industry', *Tourism Management*, 12 (1): 9–14.

Jansen-Verbeke, M. (1995) 'A regional analysis of tourist flows within Europe', *Tourism Management*, 16 (1): 73–82.

Jansen-Verbeke, M. and Dietvorst, A. (1987) 'Leisure, recreation and tourism: a geographical view on integration', *Annals of Tourism Research*, 14 (3): 361–75.

Jenkins, J.M., Hall, C.M. and Troughton, M. (1999) 'The restructuring of rural economies: rural tourism and recreation as a government response', in Butler, R., Hall, C.M. and Jenkins, J.M. (eds), *Tourism and Recreation in Rural Areas*, Chichester: John Wiley, pp. 43–67.

Jenkins, J.M. and Prin, E. (1999) 'Rural landholder attitudes: the case of public recreational access to "private" rural lands', in Butler, R., Hall, C.M. and Jenkins, J.M. (eds), *Tourism and Recreation in Rural Areas*, Chichester: John Wiley, pp. 179–96.

Jenner, P. and Smith, C. (1996) 'Attendance trends at Europe's leisure attractions', *Travel and Tourism Analyst*, 4: 72–93.

Johnson, K.A. (1990) 'Origins of tourism in the Catskill Mountains', *Journal of Cultural Geography*, 11 (1): 5–16.

Johnson, M. (1995) 'Czech and Slovak tourism: patterns, problems and prospects', *Tourism Management*, 16 (1): 21–8.

Kabanoff, B. (1982) 'Occupational and sex differences in leisure needs and leisure satisfactions', *Journal of Occupational Behaviour*, 3 (4): 233–45.

Karp, D.A. and Yoels, W.C. (1990) 'Sport and urban life', *Journal of Sport and Social Issues*, 14 (2): 77–102.

Kay, G. (1989) 'Routes for recreational walking', *Town and Country Planning*, 58 (3): 78–81.

Kay, G. (1996) *Detecting Patterns of Countryside Recreation*, Staffordshire University Department of Geography Occasional Papers, New Series A: No. 8, Stoke-on-Trent: Staffordshire University.

Kay, G. (1998) 'The right to roam – a restless ghost', *Town and Country Planning*, 67 (6/7): 255–59.

Kay, G. (2002) *Access for Countryside Walking: Politics, Provision and Need*, Stoke-on-Trent: Staffordshire University Press.
Kay, G. and Moxham, N. (1996) 'Paths for whom? Countryside access for recreational walking', *Leisure Studies*, **15** (3): 171–83.
Kelly, J.R. (1983) *Leisure Identities and Interactions*, London: George Allen and Unwin.
Kelly, J.R. (1991) 'Commodification and consciousness: an initial study', *Leisure Studies*, **10** (1): 7–18.
Knowles, T. and Garland, M. (1994) 'The strategic importance of CRSs in the airline industry', *Travel and Tourism Analyst*, **4**: 16.
Krippendorf, J. (1987) *The Holiday Makers*, Oxford: Butterworth Heinemann.
Langman, L. (1992) 'Neon cages: shopping for subjectivity', in Shields, R. (ed.), *Lifestyle Shopping: The subject of consumption*, London: Routledge, pp. 40–82.
Larkham, P. (1996) *Conservation and the City*, London: Routledge.
Lash, S. and Urry, J. (1994) *Economies of Signs and Spaces*, London: Sage.
Law, C.M. (1992) 'Urban tourism and its contribution to economic regeneration', *Urban Studies*, **29** (3/4): 599–618.
Law, C.M. (1993) *Urban Tourism: Attracting visitors to large cities*, London: Mansell.
Law, C.M. (1996) *Tourism in Major Cities*, London: International Thomson Business.
Law, C.M. (2000) 'Regenerating the city centre through leisure and tourism', *Built Environment*, **26** (2): 117–29.
Law, C.M. and Warnes, A.M. (1982) 'The destination decision in retirement migration', in Warnes, A.M. (ed.), *Geographical Perspectives on the Elderly*, Chichester: John Wiley, pp. 53–81.
Laws, E. and le Pelley, B. (2000) 'Managing complexity and change in tourism: the case of a historic city', *International Journal of Tourism Research*, **2** (4): 229–46.
Lee, Y., Dattilo, J. and Howard, D. (1994) 'The complex and dynamic nature of leisure experience', *Journal of Leisure Research*, **26** (3): 195–211.
Lehtonen, T. and Maenpää, P. (1997) 'Shopping in the East Centre mall', in Falk, P. and Campbell, C. (eds), *The Shopping Experience*, London: Sage, pp. 136–65.
Leiper, N. (1979) 'The framework of tourism: towards a definition of tourism, tourist, and the tourism industry', *Annals of Tourism Research*, **6** (3): 390–407.
Lickorish, L.J. and Jenkins, C.L. (1997) *An Introduction to Tourism*, Oxford: Butterworth Heinemann.
Light, D. and Andone, D. (1996) 'The changing geography of Romanian tourism', *Geography*, **81** (3): 193–203.
Long, P. (1999) 'Tourism development regimes in the inner city fringe: the case of "Discover Islington", London', *Journal of Sustainable Tourism*, **8** (3): 190–206.
Loverseed, H. (2001) 'Sports tourism in north America', *Travel and Tourism Analyst*, **3**: 25–42.
Lowe, P. and Ward, S. (1998a) 'Britain in Europe: Themes and issues in national environmental policy', in Lowe, P. and Ward, S. (eds), *British Environmental Policy and Europe: Politics and policy in transition*, London: Routledge, pp. 3–30.
Lowe, P. and Ward, S. (eds) (1998b) *British Environmental Policy and Europe: Politics and policy in transition*, London: Routledge.
Lowerson, J. (1995) *Sport and the English Middle Classes 1870–1914*, Manchester: Manchester University Press.
Lumsdon, L. (2000) 'Cycle tourism in Europe: EuroVelo', *Insights*, **11** (A): 143–56.

Lury, C. (1997) 'The objects of travel', in Rojek, C. and Urry, J. (eds), *Touring Cultures: Transformations of travel and theory*, London: Routledge, pp. 75–95.

Lyth, P.J. and Dierikx, M.L.J. (1994) 'From privilege to popularity: the growth of leisure air travel since 1945', *Journal of Transport History*, 15 (2): 97–116.

MacCannell, D. (1989) *The Tourist*, London: Macmillan.

MacCannell, D. (1992) *Empty Meeting Grounds: The tourist papers*, London: Routledge.

McCool, S.F. and Patterson, M.E. (2000) 'Trends in recreation, tourism and protected area planning', in Gartner, W.C. and Lime, D.W. (eds), *Trends in Outdoor Recreation, Leisure and Tourism*, Wallingford: CAB International, pp. 111–19.

Macdonald, R. (2000) 'Urban tourism: an inventory of ideas and issues', *Built Environment*, 26 (2): 90–8.

McGarvey, C. (1996) 'Access to water – irreconcilable pressures or manageable challenges?', *Countryside Recreation Network News*, 4 (3): 10–12.

McKendrick, J., Fielder, A.V. and Bradford, M.G. (2000) 'Enabling play or sustaining exclusion? Commercial playgrounds and disabled children', *North West Geographer*, 3 (2): 32–49.

McKercher, R. (1996) 'Differences between tourism and recreation in parks', *Annals of Tourism Research*, 23 (3): 563–75.

McNally, S. (2001) 'Farm diversification in England and Wales – what can we learn from the Farm Business Survey?', *Journal of Rural Studies*, 17 (2): 247–57.

MAFF (1994) *Success with Farm-based Tourist Accommodation*, London: The Ministry of Agriculture, Fisheries and Food.

Mathieson, A. and Wall, G. (1982) *Tourism: Economic, social and physical impacts*, Harlow: Longman.

May, F.B. (1983) 'Victorian and Edwardian Ilfracombe', in Walton, J.K. and Walvin, J. (eds), *Leisure in Britain 1780–1939*, Manchester: Manchester University Press, pp. 187–206.

Meethan, K. (1996) 'Place, image and power: Brighton as a resort', in Selwyn, T. (ed.), *The Tourist Image: Myths and myth-making in tourism*, Chichester: John Wiley, pp. 179–96.

Meethan, K. (2001) *Tourism in Global Society: Place, culture, consumption*, Basingstoke: Palgrave.

Miller, D., Jackson, P., Thrift, N., Holbrook, B. and Rowlands, M. (1998) *Shopping, Place and Identity*, London: Routledge.

Mintel (1996) 'Independent travel', *Leisure Intelligence*, August.

Mintel (1997a) 'Long haul holidays', *Leisure Intelligence*, March.

Mintel (1997b) 'Short breaks abroad', *Leisure Intelligence*, September.

Mintel (1997c) 'Activity holidays', *Leisure Intelligence*, October.

Mintel (1997d) 'Self-catering holidays abroad', *Leisure Intelligence*, November.

Mintel (1998) 'Inclusive tours', *Leisure Intelligence*, March.

Mintel (1999a) 'Crossing the Channel', *Leisure Intelligence*, July.

Mintel (1999b) 'Holiday centres', *Leisure Intelligence*, August.

Mintel (2000a) 'Snowsports', *Leisure Intelligence*, August.

Mintel (2000b) 'Days out', *Leisure Intelligence*, May.

Mintel (2000c) 'UK theme parks', *Leisure Intelligence*, July.

Mitchell, B.R. (1998) *International Historical Statistics: The Americas 1750–1993*, London: Macmillan Reference Ltd.

Mommaas, H. and van der Poel, H. (1989) 'Changes in economy, politics and lifestyles: an essay on the restructuring of urban leisure', in Bramham, P. *et al.* (eds), *Leisure and Urban Processes. Critical Studies of Leisure Policy in Western European Cities*, London: Routledge, pp. 254–75.

Moore, K., Cushman, G. and Simmons, D. (1995) 'Behavioural conceptualisations of tourism and leisure', *Annals of Tourism Research*, **22** (1): 67–85.

Morgan, R.H. and Jenkins, D.E. (1997) 'Rechar: too little, too late?', *Contemporary Wales*, **9**: 130–51.

Murphy, P.E. (1985) *Tourism: A community approach*, London: Routledge.

Murphy, P.E. (ed.) (1997) *Quality Management in Urban Tourism*, Chichester: John Wiley.

Nava, M. (1997) 'Modernity's disavowal: women, the city and the department store', in Falk, P. and Campbell, C. (eds), *The Shopping Experience*, London: Sage, pp. 56–91.

Newby, P. (1992) 'Shopping as leisure', in Bromley, R.D.F. and Thomas, C. (eds), *Retail Change: Contemporary issues*, London: UCL Press, pp. 208–28.

Nicholson-Lord, D. (1987) *The Greening of the Cities*, London: Routledge and Kegan Paul.

Nicholson-Lord, D. (1995) *Calling in the Country: Ecology, parks and urban life*, Park Life Working Paper No. 4, Stroud: Comedia Demos.

Olszewska, A. and Roberts, K. (eds) (1989) *Leisure and Lifestyle: A comparative analysis of free time*, London: Sage.

O'Neill, C. (1994) 'Windermere in the 1920s', *The Local Historian*, **24** (4): 217–24.

ONS (Office for National Statistics) (1998) *Living in Britain 1996 – Results from the 1996 General Household Survey*, London: The Stationery Office.

ONS (1999) *Social Trends*, No. 29, London: The Stationery Office.

ONS (2000) *Family Spending: A report on the 1999–2000 Family Expenditure Survey*, London: The Stationery Office.

ONS (2001) *Social Trends*, No. 31, London: The Stationery Office.

OPCS (Office of Population Censuses and Surveys) (1992) *Census 1991: County monitors, England and Wales*, London: HMSO.

Oppermann, M. (1999) 'Farm tourism in New Zealand', in Butler, R. *et al.* (eds), *Tourism and Recreation in Rural Areas*, Chichester: John Wiley, pp. 225–33.

Page, S.J. (1990) 'Sports arena development in the UK: its role in urban regeneration in London Docklands', *Sport Place*, **4** (1): 3–15.

Page, S.J. (1995) *Urban Tourism*, London: Routledge.

Page, S.J. (1999) *Transport and Tourism*, Harlow: Addison-Wesley Longman.

Page, S.J., Nielsen, K. and Goodenough, R. (1994) 'Managing urban parks: users' perspectives and local leisure needs in the 1990s', *Service Industries Journal*, **14** (2): 216–37.

Page, S.J. and Sinclair, T. (1992) 'The Channel Tunnel and tourism markets in the 1990s', *Travel and Tourism Analyst*, **1**: 5–32.

Parker, G. and Ravenscroft, N. (2000) 'Tourism, national parks and private lands', in Butler, R.W. and Boyd, S.W. (eds), *Tourism and National Parks: Issues and implications*, Chichester: John Wiley, pp. 95–106.

Patmore, J.A. (1983) *Recreation and Resources: Leisure patterns and leisure places*, Oxford: Blackwell.

Pearce, D.G. (1998) 'Tourist districts in Paris: structure and functions', *Tourism Management*, **19** (1): 49–65.

Pearce, D.G. and Butler, R.W. (eds) (1993) *Tourism Research: Critiques and challenges*, London: Routledge.

Pearce, P.L. (1993) 'Fundamentals of tourist motivation', in Pearce, D.G. and Butler, R.W. (eds), *Tourism Research: Critiques and challenges*, London: Routledge, pp. 113–34.

Pearlman, D.J., Dickinson, J., Miller, L. and Pearlman, J. (1999) 'The Environment Act 1995 and quiet enjoyment: implications for countryside recreation in the national parks of England and Wales, UK', *Area*, **31** (1): 59–66.
Pigram, J. (1983) *Outdoor Recreation and Resource Management*, London: Croom Helm.
Pillsbury, M. (1990) *From Boarding House to Bistro*, Boston: Unwin Hyman.
Pimlott, J.A.R. (1947) *The Englishman's Holiday: A social history*, London: Faber.
Pollard, J. and Rodriguez, R.D. (1993) 'Tourism and Torremolinos: recession or reaction to environment?', *Tourism Management*, **14** (4): 247–58.
Pollock, A. (1995) 'The impact of information technology on destination marketing', *Travel and Tourism Analyst*, **3**: 66–83.
Pompl, W. (1993) 'The liberation of European transport markets', in Pompl, W. and Lavery, P. (eds), *Tourism in Europe: Structures and developments*, Wallingford: CAB International, pp. 55–79.
Pompl, W. and Lavery, P. (eds) (1993) *Tourism in Europe: Structures and developments*, Wallingford: CAB International.
Poole, R. (1983) 'Oldham Wakes', in Walton, J.K. and Walvin, J. (eds), *Leisure in Britain 1780–1939*, Manchester: Manchester University Press, pp. 71–98.
Poon, A. (1989) 'Competitive strategies for a "new tourism"', in Cooper, C.P. (ed.), *Progress in Tourism, Recreation and Hospitality Management*, vol. 1, London: Belhaven, pp. 91–102.
Priestley, G.K., Edwards, J.A. and Coccossis, H. (eds) (1996) *Sustainable Tourism? European Experiences*, Wallingford: CAB International.
Pryce, W.T.R. (1967) 'The location and growth of holiday caravan camps in Wales, 1956–65', *Transactions of the Institute of British Geographers*, **42**: 127–52.
Ravenscroft, N. (1996) 'Access to the countryside of England and Wales: public/private partnership of the privatisation of public rights?, *Journal of Park and Recreation Administration*, **14** (1): 31–44.
Ravenscroft, N. (1999) 'Hyper-reality in the official (re)construction of leisure sites: the case of rambling', in Crouch, D. (ed.), *Leisure/Tourism Geographies*, London: Routledge, pp. 74–90.
RDC (Rural Development Commission) (1996) *The Impact of Tourism on Rural Settlements*, Rural Research Report No. 21, Cirencester: Rural Development Commission.
Redmond, G. (1991) 'Changing styles of sports tourism: industry/consumer interactions in Canada, the USA and Europe', in Sinclair, M.T. and Stabler, M.J. (eds), *The Tourism Industry: An introductory analysis*, Wallingford: CAB International, pp. 107–20.
Reid, D.G. and Mannell, R.C. (1994) 'The globalization of the economy and potential new roles for work and leisure', *Loisir et Société*, **17** (1): 251–66.
Reynolds, F. (1998) 'Environmental planning: land use and landscape policy', in Lowe, P. and Ward, S. (eds), *British Environmental Policy and Europe: Politics and policy in transition*, London: Routledge, pp. 232–43.
Ritzer, G. and Liska, A. (1997) ' "McDisneyization" and "post-tourism": complementary perspectives on contemporary tourism', in Rojek, C. and Urry, J. (eds), *Touring Cultures: Transformations of travel and theory*, London: Routledge, pp. 96–112.
Roberts, K. (1989) 'Great Britain: socio-economic polarisation and the implications for leisure', in Olszewska, A. and Roberts, K. (eds), *Leisure and Lifestyle: A comparative analysis of free time*, London: Sage, pp. 47–61.

Roberts, P. (2000) 'The evolution, definition and purpose of urban regeneration', in Roberts, P. and Sykes, H. (eds), *Urban Regeneration: A handbook*, London: Sage, pp. 9–36.
Roberts, P. and Sykes, H. (eds) (2000) *Urban Regeneration: A handbook*, London: Sage.
Roberts, R. (1983) 'The corporation as impresario: the municipal provision of entertainment in Victorian and Edwardian Bournemouth', in Walton, J.K. and Walvin, J. (eds), *Leisure in Britain 1780–1939*, Manchester: Manchester University Press, pp. 138–57.
Rojek, C. (1993) 'De-differentiation and leisure', *Loisir et Société*, 16 (1): 15–29.
Rojek, C. (1995) *Decentring Leisure: Rethinking leisure theory*, London: Sage.
Rojek, C. (1999) 'Abnormal leisure: invasive, mephitic and wild forms', *Loisir et Société*, 22 (1): 21–37.
Rojek, C. and Urry, J. (eds) (1997a) *Touring Cultures: Transformations of travel and theory*, London: Routledge.
Rojek, C. and Urry, J. (1997b) 'Transformations of travel and theory', in Rojek, C. and Urry, J. (eds), *Touring Cultures: Transformations of travel and theory*, London: Routledge, pp. 1–19.
Rose, D. (1984) 'Rethinking gentrification: beyond the uneven development of Marxist urban theory', *Environment and Planning D: Society and Space*, 1 (1): 47–74.
Ryan, C. (1991) *Recreational Tourism: A social science perspective*, London: Routledge.
Sandford, Lord (1974) *Report of the National Park Review Committee*, London: HMSO.
Scott, P. (1998) *Access to the Countryside in Selected European Countries: A review of access rights, legislation and associated arrangements in Denmark, Germany, Norway and Sweden*, Perth: Scottish Natural Heritage.
SCPR (Social and Community Planning Research) (1997) *UK Day Visits Survey – Summary Findings*, London: SCPR.
Selwyn, T. (1996) *The Tourist Image: Myths and myth making in tourism*, Chichester: John Wiley.
Sharpley, R. (1996) *Tourism and Leisure in the Countryside*, Huntingdon: Elm Publications.
Shaw, G. and Williams, A.M. (1994) *Critical Issues in Tourism: A geographical perspective*, Oxford: Blackwell.
Shaw, G. and Williams, A.M. (eds) (1997) *The Rise and Fall of British Coastal Resorts*, London: Mansell.
Shields, R. (1990) *Places on the Margin: Alternative geographies of modernity*, London: Routledge.
Shields, R. (1992a) 'The individual, consumption cultures and the fate of community', in Shields, R. (ed.), *Lifestyle Shopping: The subject of consumption*, London: Routledge, pp. 99–113.
Shields, R. (1992b) 'Spaces for the subject of consumption', in Shields, R. (ed.), *Lifestyle Shopping: The subject of consumption*, London: Routledge, pp. 1–20.
Shields, R. (ed.) (1992c) *Lifestyle Shopping: The subject of consumption*, London: Routledge.
Shoard, M. (1980) *The Theft of the Countryside*, London: Temple Smith.
Shoard, M. (1987) *This Land is Our Land: The struggle for Britain's countryside*, London: Paladin Grafton.
Shoard, M. (1999) *A Right to Roam*, Oxford: Oxford University Press.

Short, J.R., Benton, L.M., Luce, W.B. and Walton, J. (1993) 'Reconstructing the image of an industrial city', *Annals of the Association of American Geographers*, **83** (2): 207–24.

Sidaway, R. (1991) 'Marina development and coastal recreation: managing growth', *Ecos*, **12** (2): 16–22.

Siehl, G.H. (2000) 'US recreation policies since World War II', in Gartner, W.C. and Lime, D.W. (eds), *Trends in Outdoor Recreation, Leisure and Tourism*, Wallingford: CAB International, pp. 91–101.

Sinclair, M.T. and Stabler, M.J. (eds) (1991) *The Tourism Industry: An introductory analysis*, Wallingford: CAB International.

Slee, W., Farr, H. and Snowdon, P. (1997) 'The economic impact of alternative types of rural tourism', *Journal of Agricultural Economics*, **48** (2): 179–92.

Smith, C. and Jenner, P. (1995) 'Marinas in Europe', *Travel and Tourism Analyst*, **6**: 56–72.

Smith, C. and Jenner, P. (1999) 'The European ski market', *Travel and Tourism Analyst* (2): 41–64.

Smith, S.L.J. and Godbey, G.C. (1991) 'Leisure, recreation and tourism', *Annals of Tourism Research*, **18** (1): 85–100.

Smith, V.L. (ed.) (1977) *Hosts and Guests: The anthropology of tourism*, Philadelphia: University of Philadelphia Press.

Soane, J.V.N. (1993) *Fashionable Resort Regions: Their evolution and transformation*, Wallingford: CAB International.

Spink, J. (1989) 'Urban development, leisure facilities and the inner city: a case study of inner Leeds and Bradford', in Brahmam, P. *et al.* (eds), *Leisure and Urban Processes. Critical Studies of Leisure Policy in Western European Cities*, London: Routledge, pp. 195–215.

Sport England (2002) *Annual Report*, London: Sport England. Accessed via www.sport-england.gov.uk

Sports Council (1982) *Sport in the Community: The next ten years*, London: Sports Council.

Sports Council (1988) *Sport in the Community: Into the '90s*, London: Sports Council.

Sports Council (1993) *Sport in the Nineties: New horizons*, London: Sports Council.

Stebbins, R. (1982) 'Serious leisure: a conceptual statement', *Pacific Sociological Review*, **25**: 251–72.

Stebbins, R. (1997) 'Casual leisure: a conceptual statement', *Leisure Studies*, **16** (1): 17–25.

Stevens, T. (1995) 'The cultural potency of sport: a neglected heritage asset', in Fladmark, J.M. (ed.), *Sharing the Earth: Local identity in global culture*, London: Donhead, pp. 59–71.

Stillwell, J., Duke-Williams, O. and Rees, P. (1993) *The Spatial Patterns of British Migration in the Context of 1975–92 Trends*, University of Leeds School of Geography Working Paper 93/19.

Stillwell, J., Rees, P. and Boden, P. (1991) *Geographical Patterns of Migration in Britain*, University of Leeds School of Geography Working Paper No. 543.

Tanner, M. (1977) *The Recreational Use of Water-Supply Reservoirs in England and Wales*, Report No. 3, London: Water Space Amenity Commission.

Taylor, I., Evans, K. and Fraser, P. (1996) *A Tale of Two Cities: Global change, local feeling and everyday life in the north of England*, London: Routledge.

Timothy, D.J. and Butler, R.W. (1995) 'Cross-border shopping: a North American perspective', *Annals of Tourism Research*, 22 (1): 16–34.

Tobin, G.A. (1974) 'The bicycle boom of the 1890s: the development of private transportation and the birth of the modern tourist', *Journal of Popular Culture*, 7 (4): 838–49.

Torkildsen, G. (1992) *Leisure and Recreation Management*, London: E. and F.N. Spon.

Towner, J. (1996) *An Historical Geography of Recreation and Tourism in the Western World, 1540–1940*, Chichester: John Wiley.

Truitt, L.J., Teye, V.B. and Farris, M.T. (1991) 'The role of computer reservation systems: international implications for the travel industry', *Tourism Management*, 12 (1): 21–36.

Tuppen, J. (1996) 'Tourism in French cities', in Law, C.M. (ed.), *Tourism in Major Cities*, London: International Thomson Business, pp. 52–87.

Turner, J.C. and Davies, W.P. (1995) 'Farm-based tourism and recreation in the United Kingdom', *Agricultural Progress*, 70 (1): 21–43.

Turner, L. and Ash, J. (1975) *The Golden Hordes: International travel and the pleasure periphery*, London: Constable.

UNDESA (United Nations Department of Economic and Social Affairs) (2000) *Statistical Yearbook*, New York: United Nations.

United States Department of Agriculture (1962) *Report of the Outdoor Recreation Resources Review Commission*, Washington: USDA.

United States Department of Commerce (1999) *The United States National Data Book, 1999*, Austin: Hoover Business Press.

United States Department of Commerce (2000) *The United States National Data Book, 2000*, Austin: Hoover Business Press.

Urry, J. (1990a) *The Tourist Gaze: Leisure and travel in contemporary societies*, London: Sage.

Urry, J. (1990b) 'The consumption of tourism', *Sociology*, 24 (1): 23–35.

Urry, J. (1992) 'The tourist gaze and the environment', *Theory, Culture and Society*, 9 (3): 1–26.

Urry, J. (1995) *Consuming Places*, London: Routledge.

Urry, J. (1997) 'Cultural change and the seaside resort', in Shaw, G. and Williams, A.M. (eds), *The Rise and Fall of British Coastal Resorts*, London: Mansell, pp. 102–13.

Van der Borg, J. (1992) 'Tourism and the city: some guidelines for a sustainable tourism development strategy', in Briassoulis, H. and van der Straaten, J. (eds), *Tourism and the Environment: Regional, economic and policy issues*, Dordrecht: Kluwer Academic, pp. 121–31.

Vera, F. and Rippen, R. (1996) 'Decline of a Mediterranean tourist area and restructuring strategies: the Valencian region', in Priestley, G.K. *et al.* (eds), *Sustainable Tourism? European Experiences*, Wallingford: CAB International, pp. 120–36.

Vickerman, R. (1995) 'The Channel Tunnel: a progress report', *Travel and Tourism Analyst*, 3: 4–20.

Voase, R. (1995) *Tourism: The human perspective*, London: Hodder and Stoughton.

Walker, H. (1985) 'The popularisation of the Outdoor Movement, 1900–1940', *British Journal of Sports History*, 2 (2): 140–53.

Walker, S.E. and Duffield, B.S. (1983) 'Urban parks and open spaces – an overview', *Landscape Research*, 8 (2): 2–12.

Walsh, K. (1992) *The Representation of the Past: Museums and heritage in the postmodern world*, London: Routledge.

Walton, J.K. (1979) 'Railways and resort development in Victorian England: the case of Silloth', *Northern History*, **15**: 191–208.

Walton, J.K. (1983) *The English Seaside Resort: A social history*, Leicester: Leicester University Press.

Walton, J.K. (1994) 'The re-making of a popular resort: Blackpool Tower and the boom of the 1890s', *The Local Historian*, **24** (4): 194–205.

Walton, J.K. and Walvin, J. (eds) (1983) *Leisure in Britain 1780–1939*, Manchester: Manchester University Press.

Walvin, J. (1978) *Beside the Seaside: A social history of the popular seaside*, London: Allen Lane.

Ward, C. and Hardy, D. (1986) *Goodnight Campers! The History of the British Holiday Camp*, London: Mansell.

Warnes, A.M. (ed.) (1982) *Geographical Perspectives on the Elderly*, Chichester: John Wiley.

Wearing, B. and Wearing, S. (1992) 'Identity and the commodification of leisure', *Leisure Studies*, **11** (1): 3–18.

Weaver, D.B. and Fennell, D.A. (1997) 'The vacation farm sector in Saskatchewan: a profile of operations', *Tourism Management*, **18** (6): 357–66.

Welch, D. (1991) *The Management of Urban Parks*, Harlow: Longman.

Whitelegg, D. (2000) 'Going for gold: Atlanta's bid for fame', *International Journal of Urban and Regional Research*, **24** (4): 801–17.

Whitson, D. and Macintosh, D. (1996) 'The global circus: international sport, tourism and the marketing of cities', *Journal of Sport and Social Issues*, **20** (3): 278–95.

Williams, A.M. (1996) 'Mass tourism and international tour companies', in Barke, M., Towner, J. and Newton, M.T. (eds), *Tourism in Spain: Critical issues*, Wallingford: CAB International, pp. 119–35.

Williams, S. (1995) *Outdoor Recreation and the Urban Environment*, London: International Thomson Press.

Williams, S. (1998) *Tourism Geography*, London: Routledge.

Williams, S. and Jackson, G.A.M. (1985) *Recreational Use of Public Open Space in the Potteries Conurbation*, Occasional Papers in Geography No. 5, Stoke-on-Trent: North Staffordshire Polytechnic.

Williams, S. and Jackson, G.A.M. (1987) *Entertainment and Social Recreations in the Potteries Conurbation*, Occasional Papers in Geography No. 9, Stoke-on-Trent: North Staffordshire Polytechnic.

Wimbush, E. and Talbot, M. (eds) (1988) *Relative Freedoms: Women and leisure*, Milton Keynes: Open University Press.

WTO (World Tourism Organization) (1995) *Compendium of Tourist Statistics 1989–1993*, Madrid: WTO.

WTO (1997) *Tourism Market Trends: East Asia and the Pacific 1986–1996*, Madrid: WTO.

WTO (1998) *Yearbook of Tourism Statistics*, Madrid: WTO.

Index

access, 126, 144, 157–9, 184–9
   deficiencies in, 185
   effects of ownership, 186–7
   politicisation of, 130, 157, 159
acculturation, 166
activity holidays, 83–6
adventure tourism, 138
aeroplanes, 76
'aesthetic cosmopolitanism', 81–2
affluence, effects of, 10, 80, 85, 92, 107, 127, 141
agri-environmental schemes, 168, 186, 188
air charters, 11, 76
air travel, 76
Alton Towers, 46, 139
Amsterdam, 65, 89, 92, 95, 96
Anaheim, 139
Athens, 92
authenticity, 20

Baltimore, 53, 93
Bansin (Germany), 31
Barcelona, 92, 120
Bath, 27, 28, 46
behaviour, 12, 15–16, 39, 49, 83, 112–13
Benidorm, 87
Berlin, 101
Birmingham, 65, 93–4, 181
Blackpool, 31, 33, 35, 127
Bloomington (USA), 108
Boston, 127
Bournemouth, 33, 35, 37, 44
Bradford, 43
Brighton, 26, 27, 32, 37, 45
Bruges, 65
Brussels, 65
business tourism, 7, 61, 90, 93
Buxton, 28

camping, 35, 157
canoeing, 140–1
Cape May (USA), 29
caravan sites, 51, 157, 165
Cardiff, 46
car ownership, 10, 130
casual leisure, 3
Chamonix (France), 83
Channel Tunnel, 61, 63–5, 117
cinemas, 12, 44, 53, 106, 108, 113, 114
cities as tourist places, 45, 65, 87
coasts
   attitudes towards, 26
   development of, 45, 51
Colberg (Germany), 30
Cook, Thomas, 71
commodification, 3, 122, 145–7
Common Agricultural Policy, 168
communism, collapse of, 63
compatibility, of recreations, 173–5
computer information systems (CIS), 78
concert halls, 41, 95, 106, 108
conflicts, 160–2
   between users, 161
   with conservation, 161–2
   with other activity, 161
connoisseur leisure *see* serious leisure
'contested countryside', 124–5
country parks, 157, 160, 170, 171, 178, 185, 187
countryside
   basic patterns of use of, 125
   selective appeal of, 125
   values attached to, 124–5
Countryside Stewardship, 168, 186, 188–9
credit cards, 79
cross-Channel visits, growth in, 64, 117

cruising, 141
cycling, 48, 83, 106, 127–9, 131, 141, 144, 170
  in Europe, 141–3
  in France, 128–9

Dartmoor National Park, 46, 49, 132–3
Davos (Switzerland), 83
de-differentiation, 9, 18, 20–3, 189
definitive maps (of rights of way), 187
demonstration effect, 166
Doberan (Germany), 30, 31, 37
dockland redevelopment, 12, 18, 53–5, 102
  see also waterfronts

Eastbourne, 33
economic integration, 19, 147–52
eco-tourism, 42
enablement, 9–12
entertainment, popular, 40–1
Environmental Action Programmes, 167–8
European Union, 65, 79, 141, 149, 156, 167, 169
Eurostar, 64, 65
EuroVelo, 141–3
excursions, 32, 38, 39–40, 45, 48, 127, 130, 131
Exeter, 49
Exmouth, 29

fairs, 41
farming
  changes in, 148–9
  and the Common Agricultural Policy, 168
  diversification of, 149, 150, 172–3
  rationalisation of, 159
farms, as attractions, 146–7
farm tourism, 149–52
  in New Zealand, 151–2
ferry services, 64
fishing, 129–30, 140
flexible accumulation, 17
footpaths *see* public rights of way
forestry and recreation, 170
Frienberg (Austria), 143

gardening, 107
Glasgow, 65
global distribution systems (GDS), 78
globalisation, 22, 148, 180
golf, 83, 106, 131, 141, 144, 157, 165
Greenock, 32
Gulf War, 59

halls of fame, 121
Harrogate, 28
Helensburg, 32
Herinsdorf (Germany), 31
heritage coasts, 170
heritage tourism, 12, 43, 93, 102, 115, 121, 145, 165
holiday camps, 35, 51, 144
holiday villages, 144, 145, 165
holidays
  abroad, 49
  growth in number, 35
  multiple, 80–1
  with pay, 10, 35, 127
  and social class, 33, 39
'honeypots', 178–9
Hove, 33

identity and self-affirmation, 13
Ilfracombe, 33
independent travel, 82–3
information distribution systems, 78
international tourism
  growth in, 59–60, 67
  and information technology, 74
Internet, 78

land
  availability of, 157
  competition between uses, 156–7
  designation, 169–72
  ownership of, 157–9, 186–7
landscape appreciation, 126–7
Largs, 32
LEADER programme, 169
leisure
  conceptions of, 2–4
  and self-affirmation, 13
  as a state of mind, 4
  and time, 2, 17, 92
  and work, 2–3, 10, 37, 104
  and the home, 105–6, 107–8
leisure centres, 101, 108–9
leisure expenditures, 10
leisure shopping, 12, 43, 53, 66, 90, 93, 102, 104, 106, 108, 112–19
leisure time, increases in, 9–10
libraries, 11, 28, 31, 101, 104
liminal sites, 117
Linz (Austria), 143
Liverpool, 53, 165, 182
Llandrindod Wells, 28
London, 61, 64, 71, 89, 91, 92, 95, 96, 101, 109, 114, 122, 129, 130, 139, 181

Long Branch (USA), 29, 38
long distance paths, 160
long haul tourism
  growth in, 67–71
  and packages, 72–3
Los Angeles, 45, 139
Lyon, 129

Madrid, 92
Mall of America, 108, 117–19
Manchester, 32, 65, 129, 181, 182–4
market segmentation, 74, 86
marketing, 74
marinas, 51–5
migration, 37, 44–5
mobility, changes in, 10, 49, 92
motivation, 12–15, 83, 90, 131
motorways, 11, 16, 49, 75–6
museums, 95, 104, 121, 182
music halls, 41, 101

Nahant (USA), 29
national forests (USA), 135
national nature reserves, 170
national parks, 46, 49, 132–3, 135, 145, 158, 159, 162, 169, 170, 171, 177, 178
national wild and scenic rivers (USA), 172
national wilderness (USA), 172
Natural Theology, 26
Newton Abbot, 49
New York, 89, 109, 127
Norderney (Germany), 31

Olympic Games, 91, 120, 182
open access land, 186, 188
ownership, of goods, 107

package tours, 71–3, 76, 83
Paignton, 46
Paris, 61, 64, 65, 66, 89, 91, 92, 96–9, 101, 109, 127, 129, 130, 139
parks and gardens, 11, 101, 109–12, 157
partnerships, 167, 180, 184
Penrith, 144
picturesque, 27, 36, 126–7
piers, 41
place identity, 94, 110, 180
place promotion, 93, 114, 119, 184
playing fields, 102
Plymouth, 49
policy
  constraints upon, 156–60
  for integration, 170, 172–3
  for urban regeneration, 179–84

key issues in, 156–66
responses, 166–79
  in environmental improvement, 167–9
  in regional development, 169
  see also rural policy; urban policy
Portobello, 32
Port Grimaud (France), 53
Port Solent (Portsmouth), 53, 55, 181
post-industrialism, 3, 11, 17–18, 42, 85, 93, 111, 114, 124, 180
Post-Fordism, 42
Prague, 63, 92
Prestatyn, 51
Preston, 126
Princetown (Dartmoor), 178
professionalism, 73–4
public houses, 104, 106, 108
public rights of way, 187–8

rail services, 11, 16, 76, 129
  impact upon resorts, 32–3, 38
rambling clubs, 129
  see also walking
rational recreation, 27, 39, 104
RECHAR programme, 169
recreation
  definitions of, 4
  influence on resorts, 36–41
  and lifestyle, 11, 37–8, 43, 49, 114, 119
recreation grounds, 11, 101, 109, 157
recreational experience, nature of, 15
recreational impacts, 163–6
recreational tastes, 20, 43, 106, 111, 126, 127, 131
  see also tastes
restaurants, 12, 20, 41, 42, 96, 108, 109, 182, 183
Rhyl, 51
right to roam, 130, 188
Rome, 61, 89, 92
rural estates, 126, 159
rural policy, 19, 148–9
  see also policy
rural recreation
  activities, 138
  as a form of escape, 130, 131
  commodification of, 145–7
  growth of, 132–8
  impacts of industrialisation on, 126
  in the economy, 147–52
  on farms, 150–1
  selective appeal of, 135
  temporal patterns of, 135–8
  traditional patterns of, 126

INDEX

rural sport, 140–1
rural tourism
  as escape, 130
  commodification of, 145–7
  early forms of, 126–7
  growth of, 132–8
  in the economy, 147–52
  selective appeal of, 135
  temporal patterns of, 135–8

St Katharine's Dock, 53–4
St Moritz (Switzerland), 83
San Francisco, 114
Santiago de Compostella (Spain), 142
Scarborough, 27, 29
Schengen Treaty, 79
seaside resorts
  as day attractions, 49–51
  decline of, 42–51
  development of, 26–36, 127
    and hinterlands, 45–9
    and land ownership, 35
    and patronage, 28, 30
    and population growth, 26, 37
    and retirement migration, 37, 44–5
    and transport, 32–3
  distinctive aspects of, 25, 43–5
  exclusivity of, 31
  in Germany, 29–31
  and recreation, 41–56
  relation to spas, 28
  seasonal usage of, 49–50
  social differentiation of, 31
  spread of, 34, 35
seawater cures, 27, 28
second homes, 82
self-affirmation, 13
Seoul, 114
serious leisure, 3–4
Set Aside programme, 168
Sheffield, 114, 120, 129, 130, 181
shopping malls, 21, 102, 106, 108, 114, 115, 117–19, 183
short breaks, 63–7, 87, 92, 119, 120–1, 135
Sidmouth, 28
Skegness, 51
skiing, 83–5
snowboarding, 85
Southend, 28, 31, 35
spas, 27–8, 37, 127, 131
spectating, 121
sport, 104, 108–9
  as a spectacle, 119–22
  rural, 140–1

sports centres, 12
sports stadia, as attractions, 122
sport tourism, 43, 119–20
stagecoaches, 32
Staines, 139
steamships, 32, 33
Swansea, 53–4, 181
swimming baths/pools, 11, 104, 109
Swinemunde (Germany), 30, 31
Sydney, 93

taste
  for cities, 93
  in landscapes, 27, 126–7
  in recreation, 39–41
  for seaside resorts, 32
Telford, 111–12
theatres, 12, 28, 31, 40, 41, 90, 95, 101, 106
theme parks, 43, 45–6, 61, 118, 139–40, 145, 157, 182
timeshares, 82
time-space compression, 16, 92
Toronto, 122
Torquay, 33, 46, 160
Torremolinos, 87
tour operators, 72
tourism
  and behaviour see behaviour
  and business see business tourism
  definitions of, 4–5
  demand structures, 6
  experience, 15, 21, 75, 82, 87–8
  and globalisation, 22
  and heritage see heritage tourism
  impacts, 163–6
  in cities see urban tourism
  in the countryside see rural tourism
  in Eastern Europe, 62–3
  in the Mediterranean, 61–2
  and lifestyle, 11, 79–86
  and motivation, 12–13, 83, 90
    see also motivation
  and Post-Fordism, 42–3
  related problems, 62
  relationship with recreation, 1, 7–8, 13, 16, 22, 25, 39, 86–88, 90, 122, 130–1, 153, 189–90
  and shopping, 114, 115, 116–17
    see also leisure shopping
  and sport see sports tourism
  and technology, 74–9
Tourism Development Action Plans, 160
tourist city, models of, 95–6, 99–100
tourist districts, 97–8

tourist gaze, 43, 122, 126
tourist shops, 98
transport
  cross-Channel, 63–5
  effects upon resorts, 32–3
  in holidaymaking, 45–6
  in international travel, 74–6, 93
  in rural recreation/tourism, 130, 131
travel industry, 71–4
Trondheim, 142

urban governance, 181
urban policy, 18, 120
  see also policy
urban recreation, 89
  change in public-private relations, 106–7
  commercialisation of, 108
  decline in outdoor, 109–112
  development of, 101–2
  indoors, 107–9
  spatial patterning of, 102–3
urban renewal, 12, 53–5, 91, 93, 119, 120, 167, 172, 179–84
urban tourism, 65, 89
  chronologies of development, 91
  growth of, 91–2

resources, 90
spatial patterns of, 95–100, 109

Venice, 92
Vienna, 61, 92
virtual tourism, 22

wakes, 33, 38–9, 126, 127
walking, 15, 48, 83, 106, 129, 131, 144, 188
Warsaw, 92
water skiing, 141
water space
  and recreation, 173, 177–8
waterfronts, 53–5, 181
Wells, 46
West Edmonton Mall (Canada), 114, 115, 117
Weymouth, 28
White Sulphur Springs (USA), 38
Windermere, 48, 130, 145, 178
Windsor, 139
working class
  and holidays, 33
  and leisure, 38–9
World Cup (football), 120
Worthing, 28

zoning strategies, 175–8, 184